OPERATIONAL RESEARCH IN INDUSTRY

Operational Research in Industry

Edited by
Tito A. Ciriani
Stefano Gliozzi
Ellis L. Johnson
and
Roberto Tadei

Ichor Business Books
An Imprint of
Purdue University Press
West Lafayette, Indiana

First Ichor Business Book edition, 1999.

Published under license from Macmillan Press Ltd, Houndmills,
Basingstoke, Hampshire, RG21 6XS

This edition available only in the United States and Canada.

03 02 01 00 99   5 4 3 2 1

Library of Congress Cataloging-in-Publication Data applied for.

ISBN 1–55753–172–2

Printed in Great Britain

Contents

List of Figures

List of Tables

Preface: Industry Plays OR

The music has been conceived at the border of Sherwood Forest. The notes sprang from participants in the MEMIPS European Community project round table. The chief conductor gathered other composers from the AIRO'97 Conference. Global Business Intelligence Solutions of IBM offered challenging OR advanced billing. Classic tones from both European and American communities completed the score. Now four directors conduct at Macmillan Press the first performance of the 62-hand concert.

- The prelude opens with improved modeling and algorithmic instruments that contribute to playing difficult musical excerpts.
- Classic petrochemical music starts the applications of scheduling and nonlinear themes.
- Transportation industry models' variations explore the combinatorial optimization field.
- Manufacturing model composers play a new melody of popular scheduling songs.
- The sound of Horizontal Marketing and datamining gives an innovative touch.
- A search of new synthetizer experiences for the unit commitment problem closes the concert.

The following overview introduces the actual performances of OR in industry. Each chapter mainly addresses the synergetic effect of OR tools' interaction, describes the algebraic models formulated, introduces the mathematical algorithms used and reports the benefits of the implementation.

Modeling enhancements with algorithmic improvements can be applied to solve large size Mixed Integer Programming (MIP) models. In Chapter 1, Rebecca L. Ho, Ellis L. Johnson, and Tina L. Shaw (Georgia Institute of Technology, Atlanta, Georgia, USA) report their experience in using PC tools for modeling and data collection – AMPL modeling language, Microsoft Access interface and Microsoft Excel spreadsheet. The optimum solution is indicated in the original list so that the optimizer is hidden from the user.

Parallel processing seems to present one of the best opportunities for accelerating the optimization process and thereby extending their range of application. In Chapter 2 Richard S. Laundy (DASH Associates Ltd., Royal Leamington Spa, UK) describes the implementation of the parallel MIP code on networks of PCs and workstations.

Process industry plays an important role in achieving advanced results. Josef Kallrath (BASF's Mathematical Consultant Group, Ludwigshafen, Germany, and University of Florida, Gainesville, USA) solves nonlinear MIP problems in Chapter 3. They apply nonlinear MIP techniques to solve production planning in BASF's petrochemical division, tanker refinery scheduling at a refinery, and to site analysis of a large BASF factory.

Sergio Barbariol, Mauro Lusetti, Marco Mantilli and Mauro Scarioni (AGIP Petroli, Livorno, Italy, and IBM Italy, Segrate, Italy) develop in Chapter 4 an integrated system for the refinery scheduling process of lube production. It operates on vacuum, furfurolo and dewaxing, and lube production units. The MIP model addresses the implementation of set-up and scheduling of the products in order to find the best combination between sequencing and cascade.

Larry Haverly (Haverly Systems, Denville, New Jersey, USA) in Chapter 5 analyzes a petrochemical production where fourteen major cracking furnaces operate on a variety of feeds and conditions. The MIP approach applies to understanding both how the production being modeled behaves and how an LP model of the system would behave.

Turaj Tahmassebi (Unilever Research, Port Sunlight, Liverpool, UK) in Chapter 6 addresses the short-term scheduling problem for multipurpose/multiproduct fast-moving consumer goods' manufacturing plants. A continuous-time representation model for the single-stage packing system, which accommodates sequence-dependent changeovers and minimum run length, is introduced.

Transportation industry models are going to become a strategic issue for regional communication and local traffic problem solution. Uwe H. Suhl and Leena M. Suhl (Berlin and Paderborn University, Germany) in Chapter 7 discuss the solution of real-life airline fleet scheduling problems for a given flight schedule. A flight schedule determines the origin and destination of each flight, together with departure and arrival times. Each flight has to be assigned to an aircraft with sufficient capacity and has to depart within a time window.

Federico Malucelli, Maddalena Nonato and Stefano Pallottino (Milan Polytechnic, Perugia and Pisa Universities, Italy) present a solution in Chapter 8 for a flexible and low-cost public transport system. MIP models and tools to support the management of new services – the so-called Demand Responsive Transportation Systems – are discussed.

Industrial manufacturing and production operations can exploit new optimization tools with substantial economic benefits. Roberto Tadei, Federico Della Croce, and Salvatore Pucci (MRP Consulting and Turin Polytechnic, Italy) present in Chapter 9 a prototype that, using administrative and production data, produces a Gantt chart of the final production scheduling. The scheduler runs on a PC and it is based on an innovative OR and Object Oriented methodology.

Javad Ahmadi, Robert Benson, and Daniel Supernaw-Issen (Advanced Micro Devices and Distributed Software Solutions, Austin, Texas, USA) have developed a management system integrated with the supply chain. The large-size planning model in Chapter 10 aims to improve customer responsiveness and optimization techniques based on manufacturing resources.

Giorgio Fasano (Alenia Spazio, Turin, Italy) has developed a MIP model dealing with the overall cargo accommodation problem for a space carrier. The model in Chapter 11 allocates the available volume, according to modularity, accessibility, operability, static and dynamic balancing requirements, and within the logistics, physical, and functional constraints.

Innovative methodologies and optimization strategies can reshape OR applications. The new approaches can successfully cope with marketing problems. Michael P. Haydock and Eric Bibelnieks (IBM, Global Business Intelligence Solutions, Minneapolis, Minnesota, USA) introduce the Horizontal Marketing technique in Chapter 12. It allows the direct marketer to shift his focus from the product to the customer, to decrease the associated advertising cost, and to increase the profitability of future mailings.

Tito A. Ciriani, Stefano Gliozzi, and Marina Russo (IBM, Global Business Intelligence Solutions, Rome and Pisa, Italy) in Chapter 13 solve a classification problem by a new iterative LP approach based on dynamic interaction between modeling and optimizer. From a large customer's database the procedure tries to isolate the buying attitude and customer response that are the main requirements of marketing operations.

Market globalization has increased the competition of the power-generation industry. Thomas Lekane and Jacques Gheury (Tractebel Energy Engineering, Brussels, Belgium) in Chapter 14 use MIP models to determine the power generation unit's schedule. In particular, such models determine when the generating units of the system start up and shut down and define the hourly generated power. The MIP modeling approach integrates classical Lagrangian-relaxation.

Samer Takriti (IBM Research Center, Yorktown Heights, NY, USA) studies in Chapter 15 the unit commitment problem, one of the most important challenges in electric power generation. This problem refers to scheduling the generating units of an electric utility in order to meet demand constraints. Lagrangian solution methods are offered for both the deterministic and stochastic cases. A MIP model is used to refine the resulting solutions.

And now we start to play.

You can select and study your favourite pieces, or relax and enjoy the full concert.

TITO A. CIRIANI
STEFANO GLIOZZI
ELLIS L. JOHNSON
ROBERTO TADEI

Trademarks and Service Marks

CPLEX is registered trademark of ILOG Inc.

Microsoft Access is a registered trademark of Microsoft Corporation.

MIMI is a registered trademark of Chesapeake Decision Science Inc.

MOPS is a registered trademark of Prof. Dr. Uwe H. Suhl.

OMNI is a registered trademark of Haverly Systems Inc.

XPRESS-MP is a trademark of DASH Associates.

Other company, product, and service names may be trademarks or service marks of their respective owners.

Notes on the Contributors

Dr Javad Ahmadi
Technical Staff, Operations Research, Advanced Micro Devices,
  MS 6135204
E. Ben White Blvd., Austin, Texas 78741, USA
phone: +1.512-602-5825   fax: +1.512-602-7470
e-mail: jahmadi@gilon.amd.com

Dr Sergio Barbariol
IBM Italia – Petroleum Downstream Industry
Circonvallazione Idroscalo, I-20090 Segrate, Milano, Italy
phone: +39.02.5962.1

Dr Robert Benson
Technical Staff, Operations Research, Advanced Micro Devices,
One AMD Place MS 2
P.O. Box 3453, Sunnyvale, CA 94088-3453, USA
phone: +1.408-774-3843   fax: +1.408-774-3367
e-mail: rbenson@gilon.amd.com

Dr Eric Bibelnieks
Consultant for IBM's Global Business Intelligence Solutions
650 Third Avenue South, Minneapolis, MN 55402, USA
phone: +1.612-397-2611   fax: +1.612-397-2611
e-mail: ebibeln@us.ibm.com

Ing. Tito A. Ciriani
Via Mercanti 8, I-56127 Pisa, Italy
phone:+39.050.574133 [Grottaferrata: +39.06.9412297]
fax: +39.050.574133
e-mail: ciriani@cibernet.it
  [emergency e-mail: t.ciriani@agora.stm.it]

Prof. Federico Della Croce
Dipartimento di Automatica e Informatica, Politecnico di Torino,
  Italy
Corso Duca degli Abruzzi 24, I-10129 Torino
phone: +39.011. 5647059   fax: +39.011.5647099
e-mail: dellacroce@polito.it

Dr Giorgio Fasano
Alenia Aerospazio, Space Division, Turin Plant
Corso Marche 41, I-10146 Torino, Italy
phone: +39.011.7180.219   fax: +39.011.723307
e-mail: gfasano@am.to.alespazio.it

Dr Stefano Gliozzi
Consultant for IBM's Global Business Intelligence Solutions
Via Sciangai 53, I-00144 Roma, Italy
phone: +39.06.5966.5477   fax: +39.06.5966.5084
e-mail: stefano_gliozzi@it.ibm.com

Mr Jacques Gheury
Tractebel Energy Engineering, Planning Methods Section,
   Principal Engineer.
Avenue Ariane 7B, B-1200 Brussels, Belgium
phone: +32.2773.8890   fax: +32.2773.8890
e-mail: jacques.gheury@tractebel.be

Mr Larry (C.A.) Haverly
Haverly Systems Inc.
12 Hinchman Ave., Denville New Jersey 07834 USA
phone: +1.973.627.1424   fax: +1.973.625.2296
e-mail: larryh@haverly.com

Dr Michael P. Haydock
Worldwide Managing Principal for IBM's Global Business
   Intelligence Solutions
650 Third Avenue South, Minneapolis, MN 55402, USA
phone: +1.612-397-6627   fax: +1.612-397-6037
e-mail: haydock@ibm.net

Dr Rebecca L. Ho
School of Industrial & Systems Engineering
Georgia Institute of Technology, Atlanta, Georgia 30332-0205, USA
phone: +1.404.894.2316

Prof. Ellis L. Johnson
Coca-Cola Professor, School of Industrial & Systems Engineering
Georgia Institute of Technology, Atlanta, Georgia 30332-0205, USA
phone: +1.404.894.2316

e-mail: ejohnson@isye.gatech.edu
home: 1333 Council Bluff Drive NE, Atlanta, Georgia 30345, USA
   phone: +1.404.9821538

Prof. Dr Josef Kallrath
BASF-AG Aktiengesellschaft Zentralbereich Informatik
ZOI/ZC, Building C13, 67056 D-Ludwigshafen, Germany
and University of Florida, Gainesville, FL 32611, USA
phone: +49.621.60.78297   fax: +49.621.60.49463
e-mail: kallrath@zx.basf-ag.de
alternate internet routes:
   kallrath@astro.ufl.edu
   kallrath@acs.ucalgary.ca

Dr Richard S. Laundy
DASH Associates
Binswood Avenue, Royal Leamington Spa, Warwickshire CV32 5TH,
   UK
phone: +44 1926 315862   fax: +44 1926 315854
e-mail: rsl@dashcov.demon.co.uk

Mr Thomas Lekane
Tractebel Energy Engineering, Planning Methods Section Manager
Avenue Ariane 7, B-1200 Brussels
phone: +32.2773.8890   fax: +32.2773.8890
e-mail: thomas.lekane@tractebel.be

Dr Mauro Lusetti
IBM Italia – Global Services
Circonvallazione Idroscalo, I-20090 Segrate, Milano, Italy
phone: +39.02.5962.1

Dr Federico Malucelli
Dipartimento di Elettronica e Informazione,
Politecnico di Milano
Piazza L. da Vinci 32, I-20133 Milano
fax: +39 02 23993412
e-mail: malucell@elet.polimi.it

Ing. Marco Mantilli
AGIP Petroli – Raffineria di Livorno
Via Aurelia 74I, I-57017 Stagno, Livorno, Italy
phone: +39.0586.948449   fax: +39.0586.942164
e-mail: agip@etrurianet.it

Dr Maddalena Nonato
Istituto di Elettronica, Università di Perugia
Via G. Duranti 1/A1, Santa Lucia Cenetola, I-06125 Perugia, Italy
phone: +39.075.585.2685   fax: +39.075.585.2654
e-mail: nonato@istel.ing.unipg.it

Prof. Stefano Pallottino
Dipartimento di Informatica, Università di Pisa
Corso Italia 40, I-56124 Pisa
phone: +39.050.887237   fax: +39.050.887226
e-mail: pallo@di.unipi.it

Mr Salvatore Pucci
MRP Consulting
C. Ferrucci 27, I-10138 Torino, Italy
phone: +39.011.4346064   fax: +39.011.4346298

Dr Marina Russo
Consultant for IBM's, Global Business Intelligence Solutions
Via Sciangai 53, I-00144 Roma, Italy
phone: +39.06.5966.1   fax: +39.06.5966.5084
e-mail: mrusso@it.ibm.com

Ing. Mauro Scarioni
IBM Italia – Global Services
Circonvallazione Idroscalo, I-20090 Segrate, Milano, Italy
phone: +39.02.5962.5621   fax: +39.02.5962.4786
e-mail: mscarioni@vnet.ibm.com

Dr Tina L. Shaw
School of Industrial & Systems Engineering
Georgia Institute of Technology, Atlanta, Georgia 30332-0205, USA
phone: +1.404.894.2316

Prof. Dr Leena M. Suhl
Universität-GH Paderborn, Lehrstuhl für Wirtschaftsinformatik
    und OR
Warburgerstraße 100, D-33098 Paderborn, Germany
phone: +49.5251.60 3723   fax: +49.05251.60 3542
e-mail: suhl@notes.uni-paderborn.de

Prof. Dr Uwe H. Suhl
Freie Universität, Lehrstuhl für Wirtschaftsinformatik
Garystraße 21, D-14195 Berlin (Dahlem), Germany
phone: +49.30.838.5009   fax: +49.30.838.2027
e-mail: suhl@wiwiss.fu-berlin.de

Dr Daniel Supernaw-Issen
CEO Distributed Software Solutions
P.O. Box 26901, Austin, Texas 78755-0901
phone: +1.512-342-7090
e-mail: danielsi@distributed-software.com

Prof. Roberto Tadei
Dipartimento di Automatica e Informatica, Politecnico di Torino
Corso Duca degli Abruzzi 24, I-10129 Torino, Italy
phone: +39.011.5647032   fax: +39.011.5647099
e-mail: tadei@polito.it

Dr Turaj Tahmassebi
Unilever Research, Port Sunlight Laboratory
Quarry Road East, Bebington, WIRRAL, L63 3JW, UK
phone: +44.151.471.3737   fax: +44.151.471.1800
e-mail: turaj.tahmassebi@unilever.com

Dr Samer Takriti
Mathematical Sciences Department, IBM Thomas J. Watson
    Research Center
P. O. Box 218, Yorktown Heights, New York, 10598, USA
phone: +1.914.945.2458
e-mail: takriti@us.ibm.com

1 Modeling Tools for Airline-crew Scheduling and Fleet-assignment Problems

Ellis L. Johnson, Tina L. Shaw
and Rebecca L. Ho

I. INTRODUCTION

The PC and PC network environments continue to experience rapid technological advances pushing the bounds of both computing speed and memory size. Both of these factors have constrained us, in the past, from being able to solve practical optimization problems of any large size in this environment. This was especially true for problems such as airline crew scheduling and fleet assignment, whose problem size grows rapidly with the size of the input data. As the PC and PC network environments become more powerful, new tools are emerging that can harness and enable users to easily access and exploit that power.

We developed a set of PC-based modeling tools for the airline crew scheduling and fleet assignment problems. Initially, these tools were developed to support an Airline Operations Research class being taught at the Georgia Institute of Technology. The tools were designed for students and researchers to help them better understand the crew scheduling and fleet assignment problems. As such, they allow users to easily view and manipulate the model and data and see the results of the various changes they make. This usage of the tool makes it valuable in 'what-if' studies such as effects of increasing overnight rest times.

For crew scheduling the changes that can be made center on the legality rules and cost formulas used to generate the pairing (see below). It is also easy to alter the schedule, change which stations are the crew bases, and set station-specific overrides for the minimum and maximum sit and overnight rest at a particular station. For fleet assignment the changes that can be made include altering schedule, adding or

1

removing fleet types, changing the cost and revenue data, and restricting a fleet type from flying certain legs.

We developed some special-purpose code but we extensively use the PC-based applications Microsoft Access[1] (for data management and user interface) and CPLEX[2] (for optimization). For the fleet assignment problem, was also developed a PC-based AMPL[3] model.

In section 2, we discuss the modeling tools for crew scheduling. We describe our basic crew scheduling model, the applications we chose to use, the interface we developed for entering the data and problem parameters and viewing the output, data structures, and application integration. In section 3, we discuss the modeling tools for fleet assignment. We describe the underlying AMPL model, the conversion to Access/CPLEX, the data structures required, and application integration issues. (Models are available at http://tli.isye.gatech.edu/airlines/models/).

II. CREW SCHEDULING

2.1. The basic model

The basic model that we use for the crew scheduling problem is the set partitioning model. The columns or sets are called pairings. A pairing is a sequence of flight legs that represents a possible itinerary that a crew might be assigned to fly. It starts at a crew base and ends at the same crew base. It is made up of some number of duties with periods of overnight rest between the duties. Duties can be thought of as a day's work or a day's worth of flying. Figure 1.1 represents a typical pairing.

There are a significant set of rules and regulations that govern what constitutes valid duties and pairings. Such things as the time between flights, number of flights in a duty, total amount of flying time in a duty, length of duties, length of overnight rest, number of duties in a pairing,

Figure 1.1   A typical pairing

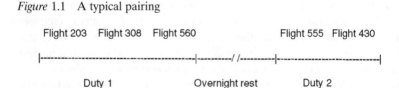

and the length of pairing are regulated. In addition, the cost of a pairing depends on several factors and, in general, can be computed only after the entire pairing has been determined. For these reasons, special-purpose algorithms are often used to generate and compute the cost of valid pairings separately from the optimizer.

In a straightforward application of the set partitioning model to crew scheduling, all of the valid pairings are generated up front. For each pairing, a leg incident vector (a vector such that each component j is a 1 if leg j is contained in the pairing and a 0 otherwise) is generated and the cost of the pairing is computed. Let a^i be the leg incident vector for pairing i and let c_i be the cost of pairing i. We define x_i as a binary variable that will be 1 if we select pairing i and 0 otherwise. The resulting set partitioning problem is

$$\min c^T x$$
$$\text{s.t } Ax = 1$$
$$x \in B^n$$

where the optimal solution is a set of pairings such that each leg is flown exactly once at the least cost. Note that with this straightforward approach, all of the complexities of the problem are dealt with in the specialized pairing generation algorithm. The optimization model is essentially fixed and quite simple conceptually. The focus of our modeling tools, for crew scheduling, is to provide the user with the ability to easily modify the various rules and regulations that drive the pairing generation algorithm and quickly see the effects on both the pairings that were generated and the ultimate solution.

Computationally, the difficulty in solving this basic crew scheduling problem is the number of columns (pairings) that may be present. For this chapter, we assume the problem is not so large that it cannot be passed to CPLEX and solved. (See Anbil *et al.*, 1992, for a global approach to solving the larger crew scheduling problems, this work expanded on the earlier sub-problem optimization discussed in Anbil *et al.*, 1991. See also Hu, 1997, for a discussion of quasi-explicit matrices for compact storage and primal–dual methods to solve the lager problems.)

2.2. Applications

Our aim was to develop a fairly graphical user interface for modifying the problem parameters. We wanted the user to be able to turn on and off various legality and cost rules and we wanted to provide easy access to default values for the parameters and help. We also knew that for the presentation of the results, especially the pairing generation results, it would be meaningful to provide 'drill-down' capability. That is, we wanted to provide the user with the ability to view a list of the generated pairings and then from any pairing 'drill down' to see the duties (or flight legs) that made up that pairing and from any duty 'drill down' to see the flight legs that made up that duty. Microsoft Access provides both the user interface and the relational tables needed to display the results.

For the pairing generation, we developed special purpose C++ code. The pairing generation code retrieves the parameter data that governs the legality checking and cost computations for the pairings. The code incorporates the parameter data into its routine for generating first the duties and then the pairings. It uses a depth first search scheme over the flight legs to enumerate all valid duties and then a depth first search scheme over the duties to enumerate all valid pairings. It outputs the results in a format that can be used by Access to display the results and provide the 'drill-down' capability above.

We chose CPLEX, which can be called from C++ programs in a PC environment quite easily, as the optimization application. We developed special-purpose C++ code that retrieves the pairings that were generated, sets up the CPLEX environment, loads the problem, and calls CPLEX to solve it. CPLEX solves the problem as a mixed integer optimization problem. Afterwards, the special-purpose code retrieves the results from CPLEX and outputs them in a format that can be used by Access.

2.3. User interface

The primary input data is the schedule. Our modeling tools assume that the user wants to solve a daily problem. That is, we assume that every flight leg in the schedule provided by the user is flown every day of the week. Often, especially for large problems, a daily problem is solved first and then weekly exceptions are incorporated into the solution. Since our primary intent was to provide users with an understanding of the basic crew scheduling problem and a forum to interactively alter various data and problem parameters, we did not feel that allowing for weekly exceptions was a high priority feature.

Therefore, the schedule data consists of only a flight leg identifier, the departure station and time, the arrival station and time and, optionally, a plane routing. The flight leg identifier need not be unique and is not used by the model. Each line item in the schedule table is assigned a unique ID and is taken as a unique leg flyable by any crew. That is, there is no advantage enjoyed when the same crew flies contiguous sections of a sequence of legs that share the same flight leg identifier (unless the plane is necessarily routed to the next flight leg in the sequence, in which case, the minimum sit is not enforced).

The departure and arrival stations should be the unique three-character airport codes. The departure and arrival times should be numbers from 0 to 2359 representing 00:00am to 11:59pm and they should be given as local times to the departure or arrival stations, respectively. The plane routing is optional. It should be a number that represents the line item in the schedule table for the leg the plane will be routed to after this leg. The tool provides an easy means to input the routing by displaying a view of the schedule showing all station activity. This is, all arrivals and departures for a station sorted by times. The tool handles all renumbering required as a result of deleting or adding to the schedule. The default value of 999 indicates that there is no routing (either no routing data was entered or the plane overnights after this leg).

An example of a line of schedule data is shown in Figure 1.2. This is flight 1155 from Boston to Chicago. It leaves Boston at 11:30am and arrives in Chicago at 1:08pm. No plane is routing is provided.

The station activity view (shown in Figure 1.3) is a very useful way to display the schedule data. In addition to making it easier to enter the routing data, it allows users an easy way to ensure that the schedule is balanced (that the number of inbound flights equals the number of outbound flights at each station). Also, it helps the user see how various connections within the pairings, and the optimal set of pairings, were made and where some schedule or parameters changes might make a difference in the pairings generated and in the ultimate optimal solution. For these reasons, the station activity view is automatically produced by the tool and made available from various screens by simply double clicking on a station code.

Figure 1.2   Line of schedule data

ID	Flight	Dep Sta	Dep Time	Arr Sta	Arr Time	Route
7	1155	BOS	1130	ORD	1308	999

Figure 1.3　Station activity view

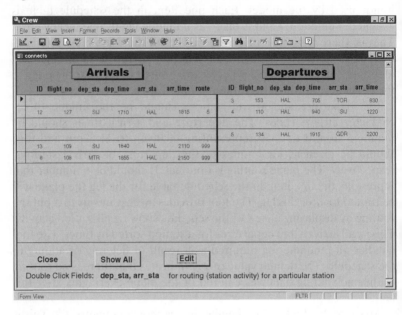

In addition to schedule data, the user provides two important pieces of information about the stations: the time zone each station is in and whether or not the station is a crew base. These data are required by the pairing generation algorithm since it must convert all times to a global standard and it must start and end pairings at only crew bases. Some station specific overrides for minimum and maximum sit and overnight rest can also be specified for any station. These values will override the parameter input screen settings.

The schedule and station data can be entered directly into the Access tables. We developed a set of easy-to-use input forms for this purpose. Schedule and station data can also be saved in a special format to text files and then later retrieved using the tool. This allows users to switch between multiple problem instances or versions of problem instances. Also, if schedule and/or station data is available from another source, it can be imported using an ASCII delimited file format.

Central to the user interface is the notion of **interactivity**. The tool is designed to allow users to easily change not only the problem data but also the problem parameters and run the pairing generation and/or the solution code again. In this way, the user can see the effects of different

Figure 1.4   Basic user interface flow

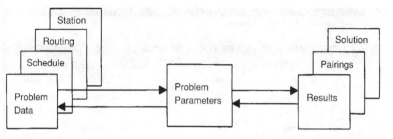

choices of rules and various data instances on the pairings that are generated and on the ultimate solution. The tool is set up so that users can move easily between problem data, problem parameter, and result screens. Figure 1.4 shows the basic user interface flow.

The **parameter input form** is at the heart of the user interface. From this form the user specifies all legality rules and cost factors that will be used by the pairing generation algorithms. All rules and cost factors can be turned on or off with check boxes. If they are turned on, a parameter can be specified (**Note**: sometimes more than one parameter is required and sometimes none is). For example, minimum overnight rest, which is the minimum amount of time required between duties, can be turned off completely or turned on and, if turned on, the user can specify the number of hours that should be used as the minimum amount of time between duties.

Default on/off settings and default parameter values are established for each installation of the tool. The user can return the entire parameter input form to the default settings or can reset any specific rule or cost factor to its default. Each rule and cost factor also has an information form associated with it. The user can request this pop-up form to receive specific information about the purpose of any rule or cost factor and how it is used by the pairing generation algorithm.

In the next paragraph, we show the parameter input form (Figure 1.5). Notice that it is divided into four sections.

1   The upper left section contains the **connection parameters**. These include minimum and maximum times for the sit between flight legs and minimum and maximum times for the overnight rest between duties.

2   The lower left section contains the **duty parameters**. These include maximum flying time and maximum number of legs per duty, maxi-

Figure 1.5   Parameter in put form

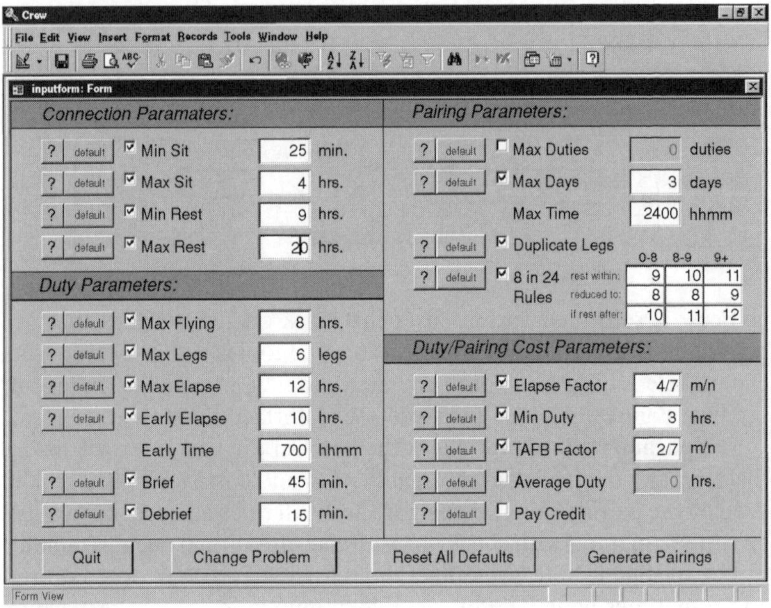

mum elapse time of a duty (and maximum elapse time of a duty beginning earlier than a certain early start time), and required brief and debrief periods before and after a duty.

3   The upper right section contains the **pairing parameters**. These include maximum number of duties in a pairing and maximum number of days in a pairing (i.e. the requirement to return by a certain time on the specified day). Also included are an indicator for whether to eliminate duplicate legs in a pairing and an indicator and related parameters for whether to impose (8-in-24) rules.

4   Lastly, the lower right section contains **cost parameters**. These include the duty elapse time factor and minimum duty guarantee used in the duty cost formula, the time away from base (TAFB) factor and the average duty guarantee used in the pairing cost formula, and an indicator for whether to compute the pairing cost as pay-in-credit or actual cost.

Once the problem data has been entered and the problem parameters specified, the user can run the pairing generation code with the click of a button. After the code executes, a form showing each station with the numbers of legs, duties, and pairings that originate from that station is displayed. From this form, users can double click on any number to see the respective legs, duties, or pairings originating from a particular station or they can click on buttons at the bottom of the screen to see all the duties or all the pairings that were generated.

The following data are included for pairings

- crew base (recall the departure and arrival station is the same and must be a crew base)
- departure day and time and arrival day and time
- TAFB (time away from base, elapse time of the pairing)
- Amount of flying time in the pairing
- cost
- number of legs in the pairing and number of duties in the pairing (see below)
- indicator for whether the pairing is in the optimal solution (included only after the solution code is run).

The following data are included for duties

- departure station, day and time and arrival station, day and time
- elapse time of the duty
- amount of flying time in the duty
- cost
- number of legs in the duty (see below).

By double clicking on the number of duties in a pairing, the user can view only those duties that are contained in the pairing. Similarly, by double clicking on the number of legs in a pairing or duty, the user can view only those legs (from the schedule) that are contained in the pairing or duty. We refer to this as being able to 'drill down' to lower levels of related information.

This 'drill-down' capability enhances the interface to the results by providing users an easy method for dissecting a pairing into its duties (or legs) and dissecting a duty into its legs. This technique provides a good visualization of the duties and pairings and how they are constructed. It also can yield valuable insights into the cost computations, the enumeration procedure, and even the data itself. And by altering the rules and rerunning the pairing generation code, users can see their changes in action (i.e. how the pairings that are generated are effected – which new ones get generated, and which current ones become illegal).

When the user is satisfied with the pairings that are generated, the

Figure 1.6 Pairing data produced

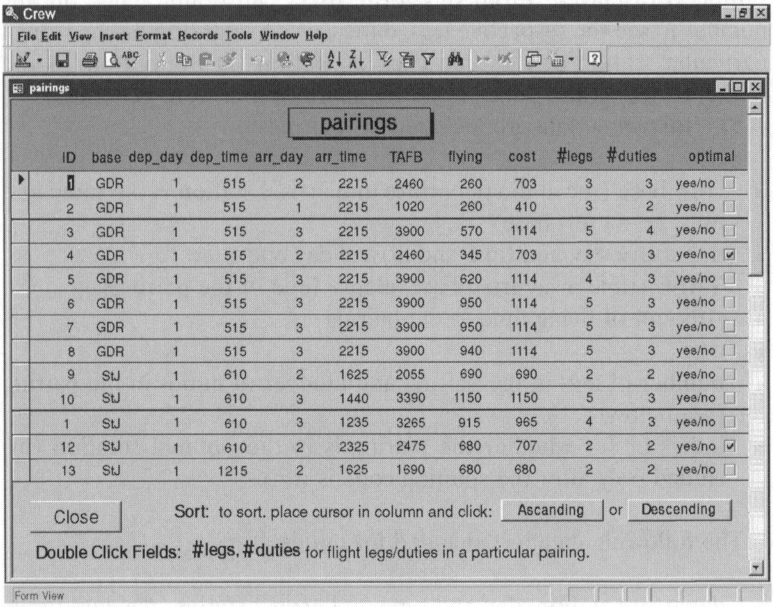

Crew

File Edit View Insert Format Records Tools Window Help

pairings

pairings

ID	base	dep_day	dep_time	arr_day	arr_time	TAFB	flying	cost	#legs	#duties	optimal
1	GDR	1	515	2	2215	2460	260	703	3	3	yes/no ☐
2	GDR	1	515	1	2215	1020	260	410	3	2	yes/no ☐
3	GDR	1	515	3	2215	3900	570	1114	5	4	yes/no ☐
4	GDR	1	515	2	2215	2460	345	703	3	3	yes/no ☑
5	GDR	1	515	3	2215	3900	620	1114	4	3	yes/no ☐
6	GDR	1	515	3	2215	3900	950	1114	5	3	yes/no ☐
7	GDR	1	515	3	2215	3900	950	1114	5	3	yes/no ☐
8	GDR	1	515	3	2215	3900	940	1114	5	3	yes/no ☐
9	StJ	1	610	2	1625	2055	690	690	2	2	yes/no ☐
10	StJ	1	610	3	1440	3390	1150	1150	5	3	yes/no ☐
1	StJ	1	610	3	1235	3265	915	965	4	3	yes/no ☐
12	StJ	1	610	2	2325	2475	680	707	2	2	yes/no ☑
13	StJ	1	1215	2	1625	1690	680	680	2	2	yes/no ☐

Close   Sort: to sort, place cursor in column and click: Ascending or Descending

Double Click Fields: #legs, #duties for flight legs/duties in a particular pairing.

Form View

problem solution code can be run. The same result screens are displayed as before but now, if an optimal integer solution was found, the pairings in the optimal solution will be identified with a check in the optimal checkbox. By sorting on this field, the user can bring all pairings in the optimal solution to the top of the list.

Figure 1.6 is an example of the pairing data produced by the tool.

2.4. Data structures

The **schedule table** is the main table used to store the problem data. Each record in the table represents a single flight leg. The tool maintains a unique leg ID, which is resequenced whenever the pairing generation code is run, as the key for the table. This resequencing keeps the indexes in line with the result tables that relate duties and pairings to legs. We discuss the result tables and their relationships in more detail later in this section.

The **stations table** stores station specific information. It is constrained by the schedule table (i.e. all stations listed as either an arrival

Figure 1.7   Connects table

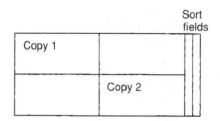

or departure station in the schedule must be included in the stations table and only those stations). The tool ensures this by controlling both the stations' view to the user and the stations outputted to the pairing generation code. A unique station ID, which is resequenced whenever the pairing generation code is run, is used as the key for the table.

The **connects table** supplies the information for the station activity view. It is derived directly from the schedule table. A sequence of queries places two copies of the schedule in one table, as shown in Figure 1.7, and loads the sort fields. The sort fields for the first half of the table are loaded with the arrival station and time while the sort fields for the second half of the table are loaded with the departure station and time. When the sort fields sort the entire table, the result is exactly the station activity view described earlier.

Whenever changes are made to the schedule table, the tool runs various queries to keep the stations an connects tables current.

There are two single record tables for the problem parameters. One stores the current parameter settings and the other stores the default settings. It is a simple matter to copy a single field from one table to the other (to restore the default setting for one parameter) or to copy the entire record from one table to the other (to restore all the default settings).

The **result tables** contain the results of the pairing generation code and the problem solution code. The pairing generation code produces four ASCII files that Access links to so that they become accessible 'tables'. These tables are duties, pairings, p2d, and d2l. The problem solution code produces one ASCII file that Access imports into a table called optimal. We discuss more about importing/exporting versus linking in section 2.5.

The result tables and related queries are designed to exploit the power of relational databases. This part of the relational data model is

Figure 1.8   Relational data model for crew scheduling results

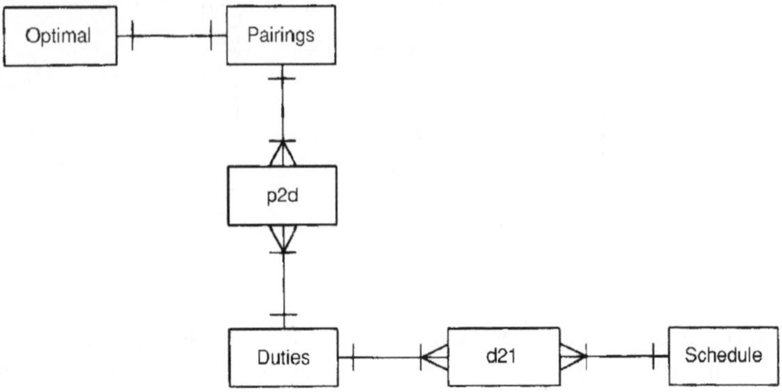

shown in Figure 1.8. The pairings and duties tables contain all the information (listed in section 2.3) about the pairings and duties that were generated. As each duty was generated, it was assigned a number starting from 1. This duty number becomes the unique ID used as the key for the duty table. Similarly, each pairing was assigned a unique ID that is used as the key for the pairings table.

The p2d table is a junction table between pairings and duties. Thus, there is a one-to-many relationship between pairings and p2d and a one-to-many relationship between duties and p2d. Each record in p2d has a unique pairing ID (P:ID) and duty ID (D:ID) combination. A particular combination appearing in this junction table indicates that pairing number P:ID contains duty number D:ID. Similarly, the d2l table is a junction table between duties and legs and each record has a unique duty ID (D:ID) and leg ID (L:ID) combination. Here, the L:ID corresponds to the leg ID in the schedule table.

In addition to identifying which duties are contained in which pairings and which legs are contained in which duties, the p2d and d2l junction tables also include data associated with each pairing–duty and duty–leg combination. These data provide sequencing and day information that are necessary to ensure that the legs/duties are displayed to the user in the proper order and with the appropriate day during a 'drill-down' request. Departures and arrivals for legs, duties, and pairings are displayed to the user as both day and time, with the time being in the local time zone.

The **optimal table** includes two fields: a pairing ID and the optimal 'yes/no' checkbox. There is a record included in the table for only those

Figure 1.9   Data flow diagram for crew scheduling

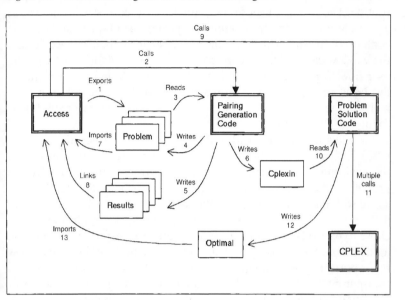

pairings that are in the current optimal solution and the checkbox is marked 'yes' for each one. In a sense, we are storing information that applies to a sub-set of pairings and thus we have a one-to-one relationship between pairings and optimal. A simple join, including all records from pairings and only those records from optimal that match, is all that is needed to correctly include the optimal checkbox field in the pairings query.

2.5. Application integration

We now discuss the integration among the various applications we used to develop the crew scheduling modeling tools. Recall that we chose Microsoft Access for the user interface. As such, Access became the primary application. It calls the special-purpose pairing generation and problem solution codes. The problem solution code, in turn, calls CPLEX.

The calls to the pairing generation code and the problem solution code are fairly straightforward (Figure 1.9). A **text file** is used to store the current status of the execution. Access sets the text file initially and then calls the appropriate special purpose code with a run application routine. The special-purpose code runs with Access, asynchronously, as a console application in a **command prompt window**. It updates the contents of the status file as it runs and, when it completes its execution, the window closes. The now visible Access status window instructs the user how to proceed. Access reads the status file into a form that is displayed to the user. The **status file** for the problem solution code includes the optimal solution value when an optimal integer solution is found.

The calls to interface with CPLEX are accomplished through the CPLEX dynamic link library routines. There are routines to set up the environment, load the problem, and instruct CPLEX to solve the problem, and retrieve status/results. We use standard CPLEX calls, so we will not discuss them in detail here.

There are data that must be shared among the applications. We chose to pass this data among the applications by exporting, importing, and, in some cases, linking to ASCII files. Specifications in Access allow the exports, imports, and links to be executed from macros, completely transparent to the user. The advantages of using ASCII files over reading/updating the Access tables directly from applications outside of Access are:

1 we can export only the data we need and in the exact format required
2 we do not need to alter existing Unix-based code that reads data from files
3 for the larger tables, where we only link to the ASCII file, it is faster and requires less memory to simply write data out to a file as the program executes
4 the resequencing and index control is easier – it becomes a direct consequence of exporting and importing when the original table is emptied prior to importing.

The data flow diagram in Figure 1.9 shows the primary data flows. The heavy outlined boxes are the applications. The light outlines boxes represent ASCII files.

When the user wants to generate pairings, Access exports the problem data: the schedule, station information, and parameters (step **1**) and calls the pairing generation code (step **2**). The pairing generation code reads the problem data (step **3**). When it completes its execution, it writes back out some of the problem data (step **4**). This is because the data may have required resequencing, and new data such as the number of duties and pairings originating from each station are needed.

The pairing generation code also writes out the four result files that Access needs to display the results (step **5**). However, these data are not arranged in a format easily used by the problem solution code, so the pairing generation code produces a version of the generated pairings that is specifically designed for input into CPLEX (step **6**). This file is called cplexin. It is a variable length file and has the following format.

Pairing ID Cost Number of Legs LegID LegID LegID ...

Access imports the new problem data into the schedule and stations tables (step **7**). However, due to the potential size of the result files and the fact that the user only accesses the results in read mode, Access links only to the result files (step **8**).

These first 8 steps may be repeated multiple times as the user tries different problem data and parameters to see the resulting duties and pairings that were generated. When the user is satisfied with the results and wants to solve the problem, Access calls the problem solution code (step **9**). The problem solution code reads the cplexin file (step **10**) and handles all the required CPLEX calls to setup, load, solve, and retrieve the solution to the problem (step **11**). Then it writes the optimal data (step **12**) which Access imports into the optimal table (step **13**).

2.6. Computation results

There are two sets of example data included with the crew scheduling model. Eastern Provincial, which is a very small schedule of only 15 flight legs, is based on an exercise in Chvatal (1983, p.352). Dixie Air, which contains 67 flight legs, is based on a 25-passenger commuter airline schedule. A modified version of the Dixie Air schedule is included with the Fleet Assignment model. The computational results given below were obtained by running the models on a Pentium 133 MHz machine with 64Mb RAM running Windows NT. The times given are approximate clock time.

For the Eastern Provincial example, we have StJ and GDR set as the crew bases. Two stations, CHA and GDR require double overnights and so the station-specific overrides for maximum rest at these two stations are set to 48 hours. With all preloaded parameter defaults, 33 pairings are generated within 1 second. The solution is also obtained within 1 second with an optimal value of 3500. This small example is ideal for interactively modifying the data/parameters and probing the resulting pairings, duties, and legs to see how they are affected by the changes.

For the Dixie Air example, we have ATL and CAE set as the crew bases. The schedule was based on turn times of 30 minutes in ATL, 24 minutes in THL, and 12 minutes at all other stations. We do not have routing data, so to be sure that the crew can follow the plane, we changed the default minimum sit to 12 and included station specific overrides of 30 for ATL and 24 for THL. We also changed the default maximum sit to 1.5 hours, the minimum rest to 9.5 hours, and the maximum rest to 20 hours. We wanted to eliminate overnight rests in ATL and THL, so we included station specific maximum rests of 9 hours (lower then the minimum rest) for these two stations. Lastly, we wanted to restrict pairings to 3 days or 3 duties which ever comes first. We did this by checking the MAX_DUTIES parameter, in addition to the MAX_DAYS parameter, and entering 3 days. With these settings, a total of 43 344 pairings were generated in 64 seconds and a solution of 4691 (312 pay-in credit, 7 percent of flying time) was obtained in 165 seconds.

III. FLEET ASSIGNMENT

3.1. AMPL model

We first formulated an AMPL model for the fleet assignment problem (Clarke, Hane, Johnson and Nemhauser, 1996). AMPL is a modeling tool that can run in the PC environment. Once a model is entered, users can load data according to the model specifications and call various optimizers to solve problems. We first describe in detail the AMPL model.

In the AMPL model, SCHED is a schedule, which contains the following information: Leg ID, Departure Station, Departure Time, Arrival Station, Arrival Time, and Equipment. Although we named it as Arrival Time in SCHED, it is actually the Ready Time, which is the arrival time plus turn time. All the time records range from 0000 to 2359. ALLTIME is a circular list of time for any combinations of stations and equipment. REDEYE, which is a subset of SCHED, has departure time greater than arrival time. Fassign is a set of binary variables. Each variable in Fassign is a record of SCHED. The solution to Fassign is 1 when we select such leg and 0 otherwise. Garc is a set of

Figure 1.10   Possible fleet types for each leg

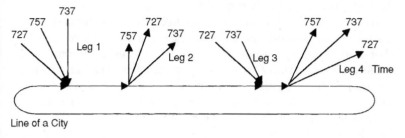

Figure 1.11   Feasible solution

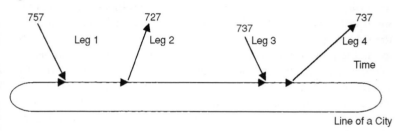

integer variables. Each variable in Garc represents a ground arc, which contains Station, Equipment, and Time. The solution to Garc is the number of planes in such ground arc.

There are three most important types of constraints in the AMPL. They are Cover, Pcount, and Balance. Cover is to make sure each leg is covered by exactly one fleet type. Pcount is for a given time. We need to make sure we don't use more planes than we have. Balance is to ensure whenever there is an in-flow, there must be a corresponding out-flow (Figure 1.10).

Figure 1.11 represents the possible fleet types for each leg. Figure 1.11 is a feasible solution after we apply the constrains.

For the complete AMPL model we used, please see Appendix 1 (p.23).

3.2. FAM in Access

AMPL is designed to be used not only as a model builder but also as a means to enter a problem instance and call an optimizer. However, we want to provide a more versatile environment for users to manipulate the model and enter and modify problem data. Once again we selected Access to provide this environment.

We took advantage of the fact that Access is a relational database. Therefore, we designed the model in the sense of relational modeling. (For a detailed description of modeling in this environment, see Atamturk *et al.*, 1996.) First we grouped variables in column strips. As mentioned above, Fassign and Garc are the two main sets of variables. Thus, our two column strips are Fassign and Garc. Col-Fassign is a table, which contains the following information: Index, Le ID, Departure Station, Departure Time, Arrival Station, Arrival Time, Fleet Type, and Net Revenue. Again, Arrival Time here is really Ready Time. Garc is a table named alltime in Access. Alltime contains the following information: Index, Station, Time, and Fleet Type. There is a slight difference between the AMPL model and the Access model. In the Access model, we added another constraint, maintenance feasibility. If the length of a ground arc is more than 5 hours, we called it a maintenance feasible ground arc. Each type of constraint is a row strip. Therefore, the row strips are Cover, Pcount, Balance, and Maintenance. Row-Cover is a table with two fields, Index and Leg ID. Row-Pcount is a table with three fields, Index, Fleet Type, and Number of Fleets. As for Balance, it is also a table name all-time. Row-Main-Sta-Eqp has three fields: Index, Station, Fleet Type. All the tables mentioned above do not allow duplicated records and all Index fields start with 1 (Figure 1.12).

Figure 1.12   The problem matrix

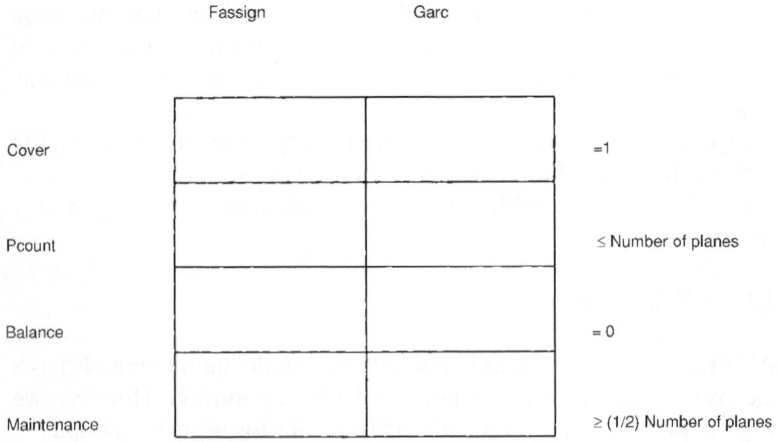

	Fassign	Garc	
Cover			=1
Pcount			$\leq$ Number of planes
Balance			= 0
Maintenance			$\geq$ (1/2) Number of planes

The rectangle in Figure 1.12 represents the problem matrix. We divided up the matrix into eight blocks (sub-matrices). Each block represents the intersection between a row strip and a column strip. All the 'block tables' have three fields, which are Row Index, Column Index and Coefficient. Usually, we use queries instead of tables (with exception of some more complicated blocks) for the 'block tables'. It is more dynamic this way because whenever we change something in a 'row table' or 'column table', queries will automatically incorporate the changes into the appropriate 'block table'. Otherwise, we would have to run a macro to make the changes every time we changed something in the 'row tables' or 'column tables'.

For each block, we find the join of a row strip and a column strip. Then, we get the Row Index from the Index of a row strip, Column Index from the Index of a column strip, and set the coefficient. For example, we get Block-Cover-Fassign when the field, Leg ID, in Col-Fassign and Row-Cover are equal. We set the coefficient to be 1. Thus, we got a table with all non-zero entries in the sub-matrix block. These are the basic techniques on how to get blocks in Access. Some of the blocks are not as easily obtainable. In that case, we break it down into different steps and then merge the tables we created together. Note that there exists no intersection between the column strip Garc and the row strip Cover. Nor does there exist an intersection between the column strip Fassign and the row strip Maintenance. Therefore, Block-Cover-Garc and Block-Main-Fassign are both zero sub-matrices.

Figure 1.13 shows what Block-Cover-Fassign looks like in Access.

After getting all the blocks, we need to offset the blocks. For example, we add the number of Fassign columns to the Column Index of Block-Pcount-Fassign, and we also add the number of Cover rows to the Row Index of the same block. After offsetting all the blocks, we merge them together into the table Result, which represents the big problem matrix we are going to pass to the solver.

Here is how we got the objective function. We used the Index in Col-Fassign as the Column Index and the corresponding Net Revenue as Coefficient for the objective function. After completing the objective function and the problem matrix, it's time to work on the right-hand side of the equations. Table Rhs has four fields: From Row Index, To Row Index, Sense, Value. When we have a record like that in Figure 1.14, it means that from row 1 to row 66 we have equation sense equal and the right-hand side value is 1. We get the Rhs table by counting up the number of rows in each row strips and sum them up in a way that will match the offset of the problem matrix.

Figure 1.13   Block-Cover-Fassign in Access

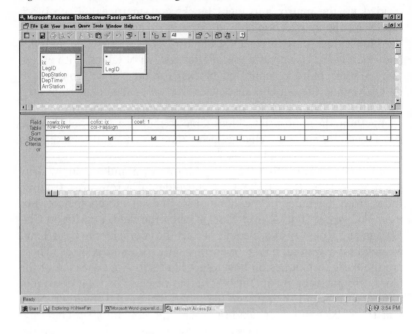

Figure 1.14   The Rhs table

FROM FOW INDEX	TO ROW INDEX	SENSE	VALUE
1	66	E	1

As before, we developed a special-purpose problem solution code to provide the interface required between the data in Access and an optimizer. And as before, we selected CPLEX as the optimizer.

3.3. Data structures

One of the main differences between the crew scheduling model and the fleet assignment model is that the crew scheduling model uses Access to store its data, but uses a C program to manipulate its data while the fleet assign model uses Access to store and manipulate its data.

There are five different sets of tables/queries in the fleet assignment model. The first set of the tables is the **instance tables**, which are provided by the user on a specific instance. These include a schedule table, a table with stations and their corresponding time zone information, a table with equipment and its turn time information, a table with stations and their turn time information, and a table with fleet types and their maintenance stations. The schedule table should contain the following information: Leg ID, Departure Station, Departure Time, Arrival Station, Arrival Time, Fleet Type, and Net Revenue. Here, all the time fields are the local time to its station and Arrival Time is the real arrival time. We don't change the departure time or arrival time until later when we run the time zone and turn time adjustment queries.

The second set of tables is the **model tables**. The tables' structures are identical to the instance tables' structures. The reason for having both instance tables and model tables is that we can create another instance and run the model again without changing anything in the model. Also, we can modify the records in the model tables without altering the instance tables.

The third set consists of **column strips, row strips, and block tables/ queries**. The fourth set is the operations queries. They consist of append queries, delete queries, make table queries, and update queries. We use these queries to manipulate data and then store the revised information into the third set of tables/queries. The third and fourth set of table/queries is the essential part of relational modeling. By using them, we are able to turn the instance into a solvable form.

The last set of tables is the **result tables**. They include the ASCII format table we pass to the solver and the ASCII format table we retrieve from as the optimal solution.

3.4. Application Integration

The application Integration for fleet assignment model is similar to the crew scheduling model, but yet simpler. Like the crew scheduling model, we used Access as the primary application. It calls the problem solution code. Then the problem solution code, in turn, calls CPLEX. Unlike the crewing scheduling model, fleet assignment model manipulates data in Access. This made the application integration simpler.

As we see in Figure 1.15, Access exports data tables including the problem matrix, right-hand side, objective function, and etc. (step **1**) and calls problem solution code (step **2**). The problem solution code

Figure 1.15   Data flow diagram for fleet assignment model

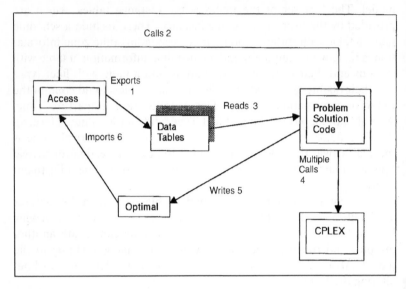

reads the data tables (step **3**) and calls CPLEX that will set up, load, and solve the problem, and then retrieve the solution to the problem (step **4**). Then, the problem solution code will write the optimal data (step **5**) which Access imports into the optimal table (step **6**).

APPENDIX 1: AMPL FORMULATION OF FLEET ASSIGNMENT MODEL

```
#leg fam

set SCHED dimension 6;
set LEGID := setof {(l, ds, dt, as. at, eqp) in SCHED} l;
set STATION := setof {l, ds, dt, as. at, eqp) in SCHED} ds;
set EQUIP := setof {l, ds, dt, as. at, eqp) in SCHED} eqp;
set STA_EQP := setof {l, ds, dt, as. at, eqp) in SCHED} (ds, eqp);
set LEG_EQP := setof {l, ds, dt, as. at, eqp) in SCHED} (l, eqp);
set REDEYE within SCHED;
set ALLTIME {STA_EQP} circular;

param cost {LEG_EQP};
param leg_rev {LEG_EQP};
param num_plane {EQUIP};

var Fassign {SCHED} >= 0;
var Garc ((s, e) in STA_EQP, t in ALLTIME[s, e]} >3 = 0;
var Revenue;
var Totcost;

maximize Profit:
    Revenue – Totcost;

subject to Addrev:
    Revenue – sum {(l,ds,dt,as,at,e) in SCHED} leg_rev[l1,e]
    *Fassign[l,ds,dt,as,at,e] = 0;

subject to Addcost:
    Totcost – sum {(l,ds,dt,as,at,e) in SCHED} cost[l,e]
    *Fassign[l,ds,dt,as,at,e] = 0;

subject to Cover {l in LEGID}:
    sum {(l,ds,dt,as,at,e) in SCHED} Fassign[l,ds,dt,as,at,e] = 1;

subject to Pcount {e in EQUIP}:
    sum {(r,ds,dt,as,at,e) in REDEYE} Fassign[r,ds,dt,as,at,e]
    +sum {(s,e) in STA_EQP} Garc [s, e, last(ALLTIME[s,e])]
    <= num_plane[e];

subject to Balance {(s,e) in STA_EQP, t in ALLTIME[s,e]}:
    Garc[s,e,t] = Garc[s,e,prev(t)]
    + sum {(l,s,t,as,at,e) in SCHED} Fassign[l,s,t,as,at,e]
    – sum {(l,ds,dt,s,t,e) in SCHED} Fassign[l,ds,dt,s,t,e] = 0;
```

Notes

1. Microsoft® Access™ is a registered trademark of Microsoft Corporation.
2. CPLEX® is a registered trademark of ILOG, Inc.
3. AMPL was developed at AT&T Bell Laboratories and is licensed through ILOG, Inc.

References

Anbil, R., R. Tanga and E. Johnson (1992) 'A Global Approach to Crew-pairing Optimization', *IBM Systems Journal*, 31(1), 71–8.

Anbil, R., E. Gelman, B. Patty and R. Tanga (1991) 'Recent Advances in Crew-Pairing Optimization at American Airlines', *Interfaces*, 2(1), 62–74.

Atamturk, A., E. Johnson, J. Linderoth and M. Savelsbergh (1996) 'ARMOS: A Relational Modeling System', Working Paper, School of Industrial and Systems Engineering, Georgia Institute of Technology.

Chvàtal, V. (1983) *Linear Programming* (New York: W.H. Freeman).

Clarke, L., C. Hane, E. Johnson and G. Nemhauser (1996) 'Maintenance and Crew Considerations in Fleet Assignment', *Transportation Science*, 30(3), 249–60.

Hu, J. (1997) 'Solving Linear Programs Using Primal-Dual Subproblem Simplex Method and Quasi-explicit Matrices', PhD thesis, School of Industrial Systems Engineering, Georgia Institute of Technology.

2 Implementation of Parallel Branch-and-bound Algorithms in XPRESS-MP

Richard S. Laundy

I. INTRODUCTION

With the recent advances in parallel computers and computer networks there has been considerable interest in developing parallel algorithms for solving combinatorial optimization problems. A natural area for research has been ways to parallelize the standard branch-and-bound (B&B) tree search algorithm for solving mixed-integer programming problems. Two approaches to parallelizing the branch-and-bound tree-search algorithm have been taken. These can be categorized as centralized storage methods and distributed storage methods (Bixby *et al.*, 1988). The centralized storage methods have a centralized storage of work packages and use a single process to distribute the work packages, while in the distributed storage methods each process has its own pool of work packages.

The code developed by Ashford *et al.* (1992) was the first parallel implementation of a commercial MIP code. This approach was a centralized scheme designed to run on an array of transputers. A similar approach was taken by Eckstein (1994) in developing a system for the CM-5. In later papers, Eckstein (1997a, 1997b) compared his earlier approach with a decentralized scheme. In the former approach, there is a theoretical lack of scalability, whilst in the latter case load balancing becomes the principal concern. Both approaches can achieve linear speed-ups on some models, however the efficiency of the tree-search algorithm can be impaired by the parallelization. In deciding whether to use a centralized or a decentralized approach to parallelize the XPRESS-MP commercial MIP code, an important consideration was how the search strategies could be maintained. These search strategies have been designed to find good solutions early on in the tree search and have been found to be effective on many models.

Ideally, we would like to develop a system which runs well on both dedicated parallel computers and on heterogeneous networks of PCs and workstations. The rationale behind this is that although dedicated parallel computers provide an excellent source of computing power, they are not always available, whereas office networks provide a cheap and readily available source of computing power which is often unused overnight and over the weekend. However, developing a system for PC and workstation networks presents a number of problems. First, the communications overhead on PC and workstation networks is a potential problem and has become increasingly important in recent years as PC and workstation processor speeds have increased rapidly whilst typical network transfer rates have remained relatively slow. Although this may not remain so in the future, the typical network transfer rate nowadays is only 10 Mbits/sec. As a result, algorithms which give good performance on dedicated parallel machines or networks of slow PCs may not give similar speed-ups on a network of machines manufactured today.

Another important consideration in designing a system for PC networks is fault tolerance. PC networks are very susceptible to failures caused by users rebooting the machines or halting processes, so it is important to be able to recover the work being performed on a processor if it fails. It is also important to be able to shut down the parallel code when the machines are needed, and restart the code from where it was halted. Typically, users would want to run the parallel code on a network overnight, stopping it in the morning, and restarting it the next night.

In this chapter we present a parallel tree-search algorithm which runs on heterogeneous networks of PCs and workstations. The parallel-search strategies mimic those of serial XPRESS-MP giving linear speed-ups and solving problems in a comparable number of nodes to the serial algorithm. We show how the communications overhead can be substantially reduced whilst providing a fault tolerant system.

II. SUMMARY OF THE BRANCH-AND-BOUND PROCESS

The branch-and-bound algorithm, first proposed by Land and Doig (1960), is a method for solving LP-models with what we call global variables – that is, variables which have discrete constraints imposed on them. We consider the case where the global variables must take on integer values, but the branch-and-bound method is equally valid for binary variables, semi-continuous variables and special ordered sets of type 1 and 2 (see Beale, 1988). That is we consider the problem:

$$
\begin{array}{ll}
\min & c^T x \\
\text{subject to} & Ax = b \\
& l \leq x \leq u \\
& x_j \in \mathcal{Z}, j \in I
\end{array}
\tag{2.1}
$$

where $c \in \mathfrak{R}^n, A \in \mathfrak{R}^{m \times n}, b \in \mathfrak{R}^m, l \in \mathfrak{R}^n, u \in \mathfrak{R}^n$ are parameters, $x \in \mathfrak{R}^n$ are the decision variables, $\mathcal{Z}$ denotes the set of integers, and I is the set of indices of the integer variables.

The branch-and-bound method is the basic method for solving mixed-integer programming problems (MIP) and is used in almost all commercial codes. Below we describe some of the node selection strategies used in XPRESS-MP.

2.1. Node selection strategies

After separating a node, a choice must be made as to which sub-problem to solve and, then, which node to select for further development. Two common strategies that are used are the best-first strategy and the depth-first strategy. In the best-first strategy, the node with the best objective function value is always selected. The advantage of this method is that the total number of nodes that has to be explored to complete the search is usually close to the minimum. However this strategy is rarely used as it stands, as the number of nodes explored before an integer solution is found can be large and before this happens the storage space for the total number of active nodes can be reached.

The depth-first strategy is better at finding solutions. In this strategy the node at the greatest depth is always selected to be solved next. As the search goes deeper, more variables are fixed to integer values, so the chances of finding an integer solution improves. However, going depth-first will not guarantee that an integer solution will be found because the nodes can become fathomed, if the sub-problems are infeasible or are cut off by a previously found integer solution.

The depth-first strategy solves the nodes quickly because there are no set-up costs when solving a sibling node immediately after solving its parent node. However, the major drawback with the strategy is that it can concentrate on parts of the tree where there are no or poor-quality solutions, thereby increasing the total number of nodes that has to be explored to complete the tree search. If an incorrect branch selection is made, then it can take a long while for the depth-first search to back-track up the tree to the correct branch.

To overcome the disadvantages of the best-first and depth-first search strategies, we use a strategy which combines the advantages of both. Central to this strategy is the calculation of estimates of the cost of branching on each fractional variable similar to those described in Benichou *et al.* (1971). At each node, up and down estimates are made of the cost of forcing each integer variable which has a fractional value, upwards to the next higher-integer value and downwards to the next lower-integer value. The sum of the minimum of the up and down estimates gives a solution estimate of the best integer solution that will be found in the sub-tree emanating from each node.

We use the solution estimates in a modified depth-first strategy to guide the search to good integer solutions. This modified depth-first strategy is a local search procedure which differs from the strict depth-first strategy in that it has a limited amount of backtracking in order to improve the chances of finding good integer solutions. First, a branch-ing variable is selected and the branch expected to lead to the better integer feasible solution is taken to create a new node. If the solution to this new node is infeasible or cut off, then the search continues from the parent node. Otherwise, the solution estimate at the new sibling node is compared with the solution estimate for the parent node. If the solution estimate of the sibling is better than that of the parent node, then the search continues from the sibling node. Otherwise, it looks like we have taken an incorrect branch and the search backtracks to the parent where the second sibling is solved. The solution estimates of the two siblings are then compared and the search continues from the sibling expected to give the best integer solution. This process contin-ues until both siblings of a parent node are fathomed. The next node to develop is then selected using the backtrack strategy. The backtrack strategy that we use is the best-first strategy, so the node with the best objective function value is selected.

By using the combined strategy outlined above, we get the advan-tages of both the best-first and depth-first strategies. The strategy selects the best outstanding node and uses this node as the starting point for a local search procedure to try to find good integer solutions. The local search is acting as a heuristic procedure removing integer infeasibilities until a solution is found or a dead end is reached. The backtrack strategy is then used to prevent the search getting stuck in one part of the tree where there may be no or poor-quality solutions.

Although the local search will not always make the correct branching decisions, it is repeated many times starting from the best outstanding node and has been found to be particularly effective in practice.

In practice, problems can range enormously in difficulty and it is essential to have a strategy which can return useful results to all types of problems. As the branch-and-bound algorithm is an NP algorithm, we must allow for a possible exponential growth in the size of the tree search which can make some problems impossible to solve to optimality. In these cases, the best we can hope to do is to find good integer feasible solutions. The strategy described above gives us a good chance of finding integer solutions if the tree search is too big to complete, whereas if the tree search can be completed, the extra number of nodes explored compared with a best-first strategy is often small, because the optimal or near-optimal integer solution is often found early in the tree search. For this reason, the combined strategy is the default strategy used in XPRESS-MP.

2.2. Variable selection strategies

The selection of variable to branch on is particularly important in the branch-and-bound algorithm if the tree search is to be completed. Improved variable selection strategies can have a substantial effect on reducing the size of the tree search. Although our parallel algorithm is independent of the variable selection strategy used, it may be advantageous to spend more time at each node selecting the variable to branch on as the increased time per node could be offset by a reduction in any communication bottlenecks.

At present, the variable selection strategy used is to calculate the minimum of the up and down estimates for each variable and select the variable with the maximum such value.

III. PARALLEL IMPLEMENTATIONS

Three general classes of parallel programming paradigms for distributed memory machines can be defined (Pritchard, 1988): farming parallelism, geometric parallelism, and algorithmic parallelism. The farming paradigm is well suited to the parallelisation of the branch-and-bound algorithm and works by decomposing the problem into many independent jobs, which are farmed out by a master process to a set of slave processes. When a slave process finishes its job, it returns the results to the master process and requests a new job. For this paradigm to work the slave processes must be able to work on the jobs allocated to them independently of the others.

Centralized storage schemes (see, for example, Ashford *et al.*, 1992; Eckstein, 1994), use the farming paradigm. In these approaches, the LPs solved at each node of the tree search are farmed out to separate slave processors. The slave processors receive the bounds and the basis for a node, solve the node and return the basis back to the master processor. With this approach, it is not possible to implement an exact depth-first or best-first strategy as it would be necessary to wait for each slave to finish to obtain the node at the greatest depth or the node with the best bound. However, it is possible to approximate these strategies by selecting only from the nodes available on the master at any point in time. By doing this the algorithm is altered and as a consequence the number of nodes required to complete the tree search can be affected. Several authors have noted these speed-up anomalies (Lai and Sahni, 1984; Li and Wah, 1986).

3.1. Parallel branch-and-bound algorithm

We now describe our scheme for parallelizing the combined search strategy described above. The combined search strategy is very difficult to implement with the farming paradigm when single nodes are farmed out to the slave processors. The reason for this is that it is difficult to coordinate the local search strategy and the backtrack strategy in a parallel environment. To get around this difficulty, we select a node on a master processor, using the backtrack strategy, and feed this node to a slave processor; the slave processor then uses the node as the starting point for the local search procedure. The local search strategy has a limited backtracking capability, so it is necessary to keep extra nodes on the slave processor to allow for this backtracking, but if the local search cannot backtrack to a node, then the node can be returned to the master processor and added to the work pool.

Initially, there is only one node (the top node) in the work pool, and this node is fed to a slave processor. As this slave processor performs its local search, nodes which can't be backtracked to are returned to the master processor, and these nodes are fed to other slave processors. This continues until all slave processors are busy, at which point, the nodes returned from the slaves are stored on the master processor. When a slave processors finishes its local search, a new node is selected from the accumulated work pool and passed to the slave processor.

To implement the local search on a slave processor, we must consider the possible states of the nodes branched on. There are four possibilities: the simplest case (case 0) occurs when the solved node is an orphan node – this is the case for the first node passed to the local search and also occurs when a parent node has been split into two sibling nodes and one of the siblings is fathomed. In this case, all that needs to be done is to select a branching variable which can be used to separate the node into two new nodes. In the remaining three cases a choice must be made about which node to select. In case 1 the preferred sibling of a parent node is solved and found to be preferable to the parent node. In this case the parent node will not be considered again in the local search and can be returned to the master. In cases 2 and 3 both sibling nodes are solved because, after solving the first sibling, the new estimates indicate that the second sibling might be preferable. In case 2 the first sibling is still selected while in case 3 the second sibling is selected. In both cases the unselected sibling can be returned to the master as it will not be considered further in the local search. These three cases are illustrated in Figure 2.1.

In terms of a family tree, nodes are returned to the master when they become grandparents or uncles. This process of returning unused nodes back to the master ensures that the master has a supply of good nodes to farm out to other slaves. Also, at any one time, a slave process has at most two active nodes (either a parent and its preferred sibling or two sibling nodes). This means that the extra storage requirements are very modest and allows the code to run on processors with limited memory and no external storage devices.

We have tried to imitate the serial strategy as closely as possible so we would expect to obtain similar node counts for most problems. However, two possible sources of speed-up anomalies are the differences in the backtrack strategy and the possibility of finding integer solutions earlier. In performing the backtrack strategy, the master processor will select from the nodes available on the master and will not have access to the nodes currently being solved on the slave processors. Although this will not make too much of a difference later in the search when there are many nodes to select from, it can have more of an effect at the start of the search when there are few nodes available on the master. If the search is large, then the overall effect of this could be beneficial if the selection of a different node leads to a good integer solution, whereas the detrimental effect is likely to be small if the effort to solve the nodes is relatively small. By performing several local searches in parallel, it can happen that the search which would be

Figure 2.1   Possible states on the slave processors

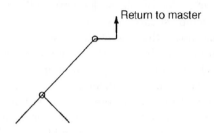

Case 1: Parent returned to master

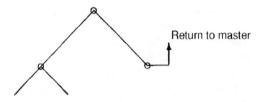

Case 2: 2nd sibling returned to master

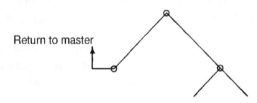

Case 3: 1st sibling returned to master

started first in the serial code is still continuing when another processor finds an integer solution which cuts off the current nodes in the first search. In this case, the first search can be terminated leading to a reduction in the total number of nodes compared with the serial code. The opposite can also happen if the search performed first by the serial code finds a good integer solution which cuts off nodes that have already been solved on other processors. In this case, the total number of nodes would increase.

Our approach is a centralized one which has a theoretical lack of scalability. As the master processor has relatively little work to perform we do not see this as too much of a problem. However, on shared memory machines where there are many available processors and relatively fast communication between the processors, it would be possible to split the central work pool between several master processes, which would then feed the nodes on to a set of slave processors under their control. The only problem would then be to ensure adequate load balancing between the master processors, which should not be too difficult.

It should be pointed out that the approach we have described does not fit entirely into the farming paradigm. As the slave processors returns nodes back to the master before the local search is complete, the work packages performed on the slaves are no longer stateless. As discussed in Section IV, this is particularly important in considering how to develop a fault-tolerant system.

IV. FAULT TOLERANCE

As we wish to run the system on an office network, it must be able to recover from processor failures. Machines can easily get turned off or crash due to other causes. Also machines may be required for other purposes and it may be necessary to stop the parallel code on a particular machine as it is taking up too much CPU time. The ability to stop and start the code is also essential. This allows the code to be run overnight or over the weekend, stopped the following morning and restarted the next night.

The interprocess communications in the system we have developed is handled by PVM (Beguelin *et al.*, 1991). Communication failures are automatically recovered by PVM and network failures can also be recovered if the network doesn't fail for too long a period. However, to deal with processor failures a mechanism is required to detect processor crashes. Fortunately there are PVM calls that check the status of a processor.

Two types of processor failure can occur: failure of the master and failure of a slave. The failure of the master is a complete failure and we treat this as unrecoverable. A checkpoint system for implementing fault tolerance on the master has been proposed in Silva *et al.* (1993), but this not straightforward to implement in our case.

It is usually easy to recover from the failure of a slave in the farming paradigm since the slaves are usually stateless processes. If this is the case, all that needs to be done is to keep a record of the job sent to a slave, and if the slave fails, re-broadcast the job to another slave. However, in our case the slaves are no longer stateless as they return nodes to the master as the local search goes deeper down the tree. Here we propose a mechanism for working out the state of a slave based on the nodes returned to the master.

To start with, the master broadcasts a node to a slave and a record is kept of this information. The master then receives a series of nodes back from the slaves. Sections 1–3 of Figure 2.1 show the three possible situations. If a parent node is returned to the master (Case 1) then this can be used to determine the current state of the slave. On the other hand if a sibling node is returned to the master (Cases 2 and 3) then the branching variable of the parent node can be obtained and the state of the slave kept on the master can be updated. Note that the slave state cannot be determined by swapping the branching direction of the parent branching variable in the returned sibling node as other variables may have been fixed in the sibling node by reduced cost fixing.

When developing the parallel code, a faulty connection in the network resulted in frequent failures of one of the machines. This situation was handled well by the fault tolerance described here. Another interesting feature of the fault tolerance arose when a bug in an early version of the code caused one of the slave processors to fail at a particular node. The master duly recovered the rogue node and sent it out to the next node which in turn failed. This continued until all the nodes failed.

V. COMPUTATIONAL RESULTS

In implementing the parallel version of XPRESS-MP it was necessary to redesign the branch-and-bound harness used within XPRESS-MP. It was necessary to split the harness into two main routines. The first main routine implements the backtrack strategy by selecting a node to develop further from all active nodes. The selected node is then passed to the second routine which performs the local search. Once this redesign was completed, the parallelization of the code was relatively straightforward.

We have run the parallel code using up to four 200 MHz COMPAQ Pentium Pros as slave computers and a 133 MHz Pentium as the master computer. The PCs were linked together with a 10 MBits/sec ethernet network. This is fairly slow but typical of the speed of an office network. To reduce the communication times, we use a high-performance version of PVM (HPPVM) which runs under Windows 3.1, 95 and NT.

Table 2.1   Test problems from the MIPLIB test suite

Matrix	Rows	Columns	Nonzeros	Integer
air04	823	8904	72965	8904
air05	426	7195	26402	7195
Cap6000	2176	6000	48249	6000
1152lav	97	1989	9922	1989
Misc07	212	260	8619	259
mod011	4480	10958	22254	96
nw04	37	87482	724148	87482
Qiu	1192	840	3432	48

Table 2.2   LP and optimal solutions for test problems

Matrix	LP solution	Optimal solution
air04	55535.43639	56137.0
air05	25877.60927	26374.0
Cap6000	−2451537.325	−2451403.234
1152lav	4656.363636	4722.0
Misc07	1415.0	2810.0
mod011	−62081950.29	−54558535.01
Qiu	−931.638857	−132.873137

The code was run on a number of the harder problems from the MIPLIB (Bixby *et al.*, 1988) test suite. Table 2.1 shows the sizes of the problems and Table 2.2 shows the objective function of the initial LP after presolve and the optimal integer solution value. In some cases, the presolve strengthens the LP relaxation by fixing integer variables and altering matrix coefficients, so the duality gap is often reduced by this procedure.

Table 2.3 shows the timings for the serial code on a 200 MHz COMPAQ Pentium Pro and the elapsed time for the parallel code on up to 4 slaves. Timings exclude solution of initial node but include time to set up slave processes and broadcast copies of the MIP to them. The runs were done with the all the XPRESS-MP options set to their default values.

Comparison of the serial times and the one-slave times in Table 2.3 shows the communications overhead involved in passing the work

Table 2.3   Serial times and elapsed times on up to four slave processors

Matrix	Serial	1 slave	2 slaves	3 slaves	4 slaves
air04	8340	8459	3409	2593	2072
air05	15071	15234	6604	5435	3993
Cap6000	782	790	156	56	49
1152lav	278	341	208	135	118
Misc07	666	1029	519	329	239
mod011	13208	13848	7030	4678	3485
nw04	6601	6732	3432	2468	1759
Qiu	4548	4958	2504	1647	1261

Table 2.4   Nodes taken by serial code and nodes on master processor for up
to four slave processors

Matrix	Serial	1 slave	2 slaves	3 slaves	4 slaves
air04	1807	464	384	432	449
air05	15105	3713	3312	3934	3862
Cap6000	7478	6720	5030	3886	3668
1152lav	11213	2662	3287	3220	3602
Misc07	60169	14341	14397	14282	14208
mod011	62892	19667	19681	19680	19682
nw04	3907	1122	1158	1078	1191
Qiu	37192	10307	10587	10548	10532

packages to and from a slave computer. The increase in the solution times are significant only on the smaller problems misc07, 1152lav and Qiu being 54.5 percent, 22 percent and 9 percent. These are relatively small problems with less than 10000 matrix elements. The average times to solve a node is as low as 0.025 seconds/node for 1152lav.

Table 2.4 shows the number of nodes in the serial case and the number of nodes returned to the master in the parallel runs. The difference between the serial and parallel nodes shows the reduction in the communications overhead by performing the local search on the slave processors. If a node is fathomed on a slave processor, is not be returned to the master processor. Also if both siblings are solved on a slave processor the parent node is not returned to the master. If every node were returned to the master then the number of nodes taken by the serial code and the parallel code with one slave would be identical.

As can be seen the number of nodes processed by the master is often reduced by a factor of 4.

The number of tree nodes processed by the master is fairly consistent between the different number of processors. One exception is the Cap6000 matrix. This is an interesting matrix in that the serial code selects a wrong branch early on in the tree search and the selected path continues to a depth of around 2000 before it terminates, whereas the path to the optimal solution is fairly short. In an early version of the parallel code, the slave processors received only a cut-off found by another processor when they completed their local searches and required a new node. This meant that when solving the Cap6000 matrix with more than one slave processor, the local search which went to a depth of around 2000 would continue for a long while after the rest of the tree search had been completed. To get around this problem, we found it necessary to broadcast improved cut-offs to all the slaves as soon as they are found. When a slave completes the solution of a node, it probes the communications channel to see if a cut-off or a request to stop has been sent by the master.

Figure 2.2 shows the speed-ups that are obtained for the test problems with the different number of processors. The problems air04 and air05 give approximately linear speed-ups – sometimes superlinear speed-up is obtained if a good integer solution is found earlier in the tree search. The problem Cap6000 shows a marked superlinear speed-up with a speed-up of over 16 with four processors. The reason, as described above, is due to finding the optimal integer solution much earlier in the tree search, but this superlinear speed-up is not likely to continue above four processors. The time taken to solve the sub-problems for the problem 1152lav is relatively small and only a speed-up of 3 can be obtained with four processors. For the remaining problems very close to linear speed-ups are obtained.

All the MIPLIB tests were run with the XPRESS-MP parameters set to their defaults. For some of the models, it is possible to reduce the size of the tree search by using priorities to select the order in which variables are branched on. For example, by using priorities, the air04 matrix can be solved in 544 seconds (163 nodes) and the air05 matrix can be solved in 288 seconds (168 nodes) using four COMPAQ Pentium Pro processors.

To test the code further we ran the code on two industrial problems (Table 2.5). The first, gas285, is an energy planning problem which had never been solved before, and the second, csusemc, is a resource planning problem developed at BASF. The gas285 problem had been so

Figure 2.2   Speed-ups for up to four slave processors

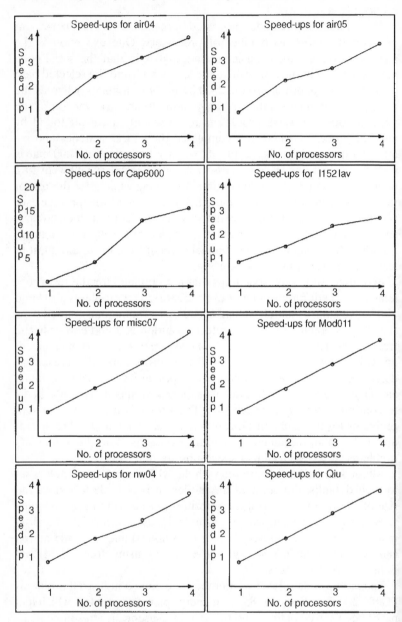

Table 2.5   Industrial test problems

Matrix	Rows	Columns	Nonzeros	Integer	Sets	Optimal solution
gas285	27528	15189	76953	1104	0	4718.219856
csusemc	9262	9446	85227	7660	492	34405.99

Table 2.6   Serial times and elapsed times on up to four slave processors

Matrix	Serial	1 slave	2 slaves	3 slaves	4 slaves	8 slaves
gas285	36355	39894	20271	13506	10134	5413
csusemc	23983	24601	9823	7216	5422	3945

Table 2.7   Nodes taken by serial code and nodes on master processor for up to four slave processors

Matrix	Serial	1 slave	2 slaves	3 slaves	4 slaves	8 slaves
gas285	20035	7782	7634	7623	7476	8044
csusemc	2704	1152	913	1026	1010	1263

difficult to solve that an alternative formulation had been devised which simplified the problem by approximated parts of it. The eventual solution of the model helped to confirm the accuracy of this simplified model. In order to solve both these problems, it was necessary to devise priorities on the branching variables in order to reduce the size of the tree search. With these priorities, the solution times are shown in Table 2.6. The eight slave times are for a network of four COMPAQ Pentium Pros and a four-processor Intel box, which has a comparable speed per processor as a single COMPAQ machine. On these difficult problems, linear or superlinear speed-ups are obtained for up to four processors and a slightly less than linear speed-up is obtained for eight processors. The number of nodes in the tree search, shown in Table 2.7, is fairly constant for different numbers of processors and the reduction in the number of nodes that have to be returned to the master as a result of performing the local searches on the slave processors is about 2.5.

VI. CONCLUSIONS

We have shown how the branch-and-bound algorithms within XPRESS-MP can be parallelized to produce linear speed-ups even on modestly sized problems. As a general rule, if the time to solve a node is greater than 0.1 seconds/node, then linear speed-ups can be expected for up to four slave processors running on a typical office network. For problems with nodes taking considerably longer than 0.1 seconds/node linear speed-ups can be expected for many more processors.

As the search strategies we have implemented are analogous to the serial strategies, similar behavior is observed with the parallel and serial codes. An exception to this occurs when the parallel code finds better solutions early on in the search thereby allowing nodes that previously were solved to be cut off. Another advantage of copying the serial strategies is that if the tree search cannot be completed the parallel code is likely to find good integer solutions.

The system we have developed is fault tolerant, which is essential for a system running on an office network as machines can easily be turned off or crash. The system also has the advantage that it can be run overnight, stopped in the morning and restarted the next night. This gives the possibility of reducing the gap between the best bound and the best solution or even prove optimality on extremely hard problems by using unused resources.

We have proposed a centralized storage scheme for the work packages, but we have also indicated how the approach can be extended to use several master processors if the communication between master and slaves becomes a bottleneck.

Future work will involve extending the parallel code to include cut generation routines. A branch-and-cut code has been developed using the XPRESS-MP sub-routine library (Cordier *et al.*, 1997), and has been found to be very successful for a number of problems. Integrating these routines into parallel XPRESS-MP pose a number of problems but are likely to parallelize well.

The results show that substantial speed improvements can be achieved on real industrial problems.

References

Ashford, R.W., P. Connard and R.C. Daniel (1992) 'Experiments in Solving Mixed Integer Programming Problems on a Small Array of Transputers', *Journal of the Operational Research Society*, 43, 519–31.

Beale, E.M.L. (1988) *Introduction to Optimization* (New York: Wiley).

Beguelin, A., J. Dongarra, A. Geist, R. Manchek and V. Sunderam (1991) 'A Users' Guide to PVM Parallel Virtual Machine', *Technical Report*, ORNL/ TM11826, Engineering Physics and Mathematics Division, Oak Ridge National Laboratory, Oak Ridge, TN.

Bixby, R.E., S. Ceria, C.M. McZeal and M.W.P. Savelsbergh (1988) An updated mixed integer programming library: Miplib 3.0. (available at: http:// www.caam.rice.edu.)

Cordier, C., H. Marchand, R.S. Laundy and L.A. Wolsey (1997) 'bc-opt: A branch-and-cut code for mixed integer programs', *Technical Report*, CORE7 CORE, Université Catholique de Louvain, Louvain-la-Neuve.

Eckstein, J. (1994) 'Parallel Branch-and-Bound Algorithms for General Mixed Integer Programming on the CM-5', *SIAM Journal on Optimation*, 4, 794–814.

Eckstein, J. (1997a) 'Distributed versus Centralized Storage and Control for Parallel Branch and Bound: Mixed Integer Programming on the CM-5', *Computational Optimization and Applications*, 7, 199–220.

Eckstein, J. (1997b) 'How Much Communication does Parallel Branch and Bound Need?', *ORSA Journal on Computing*, 9, 15–29.

Gendron, B. and T.G. Crainic (1994) 'Parallel Branch-and-Bound Algorithms: Survey and Synthesis', *Operational Research*, 42, 1042–66.

Lai, T.-H. and S. Sahni (1984) 'Anomalies in Parallel Branch-and-Bound Algorithms', *Communications of the Association for Computing Machinery*, 27, 594–602.

Land, A.H. and A.G. Doig (1960) 'An Automatic Method for Solving Discrete Programming Problems', *Econometrica*, 28, 497–520.

Li, G.-J. and B.W. Wah (1986) Computational Efficiency of Parallel Combinatorial OR-tree Searches', *JEEE Transactions, Software Engineering*, C-35, 568–73.

Pritchard, D.J. (1988) *OUG-7 Parallel Programming on Transputer-Based Machines* (Amsterdam: IOS Press).

Silva, L.M., B. Veer and G.S. Silva (1993) 'How to Get a Fault Tolerant Farm', in R. Grebe *et al.* (eds), *Transputer Applications and Systems '93* (Amsterdam: IOS Press), 923–38.

Benichou, M., J.M. Gauthier, P. Girodet, G. Hentges, G. Ribière and O. Vincent (1971) 'Experiments in Mixed-Integer *LP*', *Mathematical Programming*, 1, 76–94.

3 Mixed-Integer Nonlinear Programming Applications

Josef Kallrath

I. INTRODUCTION

In this chapter was apply different approaches to solve four rather different MINLP problems: special extensions to time-indexed formulations of production planning problems; a production planning problem in BASF's petrochemical division; a site analysis of one of BASF's bigger sites; and a process design problem. The first problem is related to a useful nonlinear extension of production planning problems based on time-indexed formulations. The second problem[1] leads to a mixed-integer nonlinear model for describing a petrochemical network including several steam crackers and plants located at two different sites. The third problem, a network design problem, leads to mixed-integer nonlinear programming problem dominated by pooling problems (Kallrath and Wilson, 1997, hereafter KW97, Section 11.1.2), as is true for the second problem. The pooling problem refers to the intrinsic nonlinear problem of forcing the same (unknown) fractional composition of multi-component streams emerging from a pool (e.g. a tank or a splitter in a petrochemical network). The fourth problem is concerned with a process design problem in which some process parameters and the topology of a system of chemical reactors are the degrees of freedom to optimize total production, selectivity, energy, and costs. All numerical experiments and production runs have been carried out on a 166 MHz Pentium processor.

The covered models lead to mixed-integer nonlinear programming (MINLP) – i.e. to optimization problems of the form

$$\min\left\{ f(\mathbf{x},\mathbf{y}) \middle| \begin{array}{ll} \mathbf{g}(\mathbf{x},\mathbf{y})=0 & \mathbf{x}\in \mathbb{R}^{n_c} \\ \mathbf{h}(\mathbf{x},\mathbf{y})\geq 0 & \mathbf{y}\in \mathbb{Z}^{n_d} \end{array} \right\}, \tag{3.1}$$

with n_c continuous and n_d discrete variables, nonlinear objective function $f(\mathbf{x}, \mathbf{y})$ and constraints $\mathbf{g}(\mathbf{x}, \mathbf{y})$ and $\mathbf{h}(\mathbf{x}, \mathbf{y})$. Problems such as

(3.1) are very difficult to solve. They belong to the class of $\mathcal{NP}$-complete problems. An overview of algorithms capable of solving such problems is given in Leyffer (1993), Floudas (1995) or KW97 (Section 11.4).

Many mixed-integer linear programming (MILP) problems are combinatorial optimization problems for which the branch and bound algorithm (Nemhauser and Wolsey, 1988) based on linear programming (LP)-relaxation proves sufficiently efficient. The algorithm is deterministic but in the worst case we see complexity growing exponentially in the problem size. Nonlinear programming (NLP) problems force us to distinguish between local and global optima. Algorithms to solve NLP problems (see, for instance, Bazaraa *et al.*, 1993) have their roots in calculus and depend on the concept of convergence. Except in special cases it is not possible to prove that an NLP algorithm converges to the global optimum. So we should keep in mind that (a) in nonconvex NLP or MINLP problems we cannot strictly prove optimality or provide safe bounds, and (b) that if nonlinear equations are present the NLP or MINLP is immediately nonconvex.

Unfortunately, MINLP problems combine the difficulties of both its sub-classes: MILP and NLP. Even worse, in addition they have properties absent in NLP or MILP. While for convex NLP problems a local minimum is identical to the global minimum, we find that this result does not hold for MINLP problems. The best we can do in nonconvex MINLP is to provide a safe bound or relative optimality with respect to a certain local optimum of the continuous problem.

II. NONLINEAR EXTENSIONS TO PRODUCTION NETWORK MODELS

This section describes an extension which can be added to any planning model based on time-indexed formulations with at most one set-up change per period. Here we apply this approach to a special case and extend the production planning problem and its model (M1, for brevity) described in KW97 (Section 10.4). As already discussed in KW97 (Chapter 6) certain nonlinear terms (absolute value terms, products of binary variables) can be expressed by linear relations involving additional binary variables. The key idea used in the current production planning problem is to replace products of continuous variables and binaries, theta-functions and absolute value terms by linear relations involving additional binary variables.

2.1. Batch constraints across periods

The motivation for the model approach developed in Section 2.2 has its root in batch or campaign production in the chemical process industry. Batch production operates in integer multiples of batches where a batch is the smallest unit to be produced, e.g. 200 tons. Several batches following each other immediately establish a campaign. Some typical batch restrictions group batches into campaigns, or consider that only campaigns of a minimal size can be produced. The batch reactors can be, for example, operated in different modes producing several products in each mode with different free or fixed recipes leading to a general mode–product relation (KW97, 153–5, 320–4): in a certain mode several products are produced (with different daily production capacity rates) and, conversely, a product can be produced in different modes. Daily production can be less then the capacity rates. Within a fixed planning horizon, T, a certain product can be produced in several campaigns. In the context of time-indexed formulation where variables p_{pt} describe the production [e.g. in tons] of a product p in period (time-interval) t it is not easy to model such batch restrictions if the batch or minimal campaign size is larger than the capacity per period. Assume that production is performed in batches of 200 tons, and that our time intervals have a length of 10 days with a daily production rate of 10 tons/day. The minimum time to produce the batch would cover 20 days, or exactly two time intervals. A plan looking like $p_{p4} = 45$ tons, $p_{p5} = 100$ tons, and $p_{p6} = 55$ tons covers three periods (the first and third only partial) to produce exactly 200 tons, and thus provides more degrees of freedom. Brockmüller and Wolsey (1995) solved the problem for a special case (production equals the capacity rates). Their approach uses explicitly the feature the production equals the capacity rates in order to compute *a priori* the number of periods to produce a campaign of specified minimal size. If daily production can take any value between zero and the capacity rate, or if a product is produced, for example, according to general mode–product relation, then this *a priori* information is not available. Our approach does not depend on this *a priori* information and can be used for more general cases.

2.2. Formulation of batch constraints

Our goal is to compute the amount, p_{rpnt}^{C}, of product $p \in \mathcal{P}$ produced for a certain campaign n in period $t \in \mathcal{T}$. The mathematical model is, for a certain site, unit, or reactor $r \in \mathcal{R}$, based on some binary state variables δ^P_{rpt} indicating whether product p is produced on r in period t, and binary start-up variables δ^S_{rpt} indicating whether the production of p is started in period t on r. Let P^-_{rpt} and P^+_{rpt} be bounds on p_{rpt} if $p_{rpt} > 0$. We may choose the upper bound P^+_{rpt} for p_{rpt} – e.g. as the length of the period (in days) times the daily production capacity – and the lower conditional bound $P^-_{rpt} = 0.8P^+_{rpt}$.

Let us, at first connect δ^P_{rpt} to the production variables p_{rpt} starting with the inequalities

$$p_{rpt} \le P^+_{rpt}\delta^P_{rpt}, \quad \forall\{rpt\} \tag{3.2}$$

If Δ_{rp} tells us whether product p is produced at the beginning of the first period, and $\Sigma_p\Delta_{rp} = 1$, then for the first period we have

$$P^-_{rp1}\delta^P_{rp1} - P^+_{rp1}(1-\Delta_{rp}) - P^+_{rp1}(1-\delta^P_{rp2}) \le p_{rp1}, \quad \forall\{rp\} \tag{3.3}$$

and for all other periods (except the last one) $T_1 := \{2, \ldots, T-1\}$

$$P^-_{rpt}\delta^P_{rpt} - P^+_{rpt}\delta^S_{rpt} - P^+_{rpt}(1-\delta^P_{rpt+1}) \le p_{rpt}, \quad \forall\{rpt \in T_1\} \tag{3.4}$$

The inequalities (3.2)–(3.4) hold the *positivity conditions* ($\delta^P_{rpt} = 0 \Leftrightarrow p_{rpt} = 0$) and ($\delta^+_{rpt} = 1 \Leftrightarrow P^-_{rpt} \le p_{rpt} \le P^+_{rpt}$) for all inner periods of a compaign. The second and third term on the left-hand side of (3.4) ensure that the positivity conditions is not applied to the first and last period of campaigns.

Now the need to relate the start-up variables to the state variables. This part depends on the problem considered. A formulation, valid for any continuous variable (e.g. the production variable p_{rpt} or the variable m^D_{rmt} denoting the time spent in mode m both used in M1) subject to constraints cross-periods, and the conditions that we can produce only one product per time and that at most two products can be produced during one period (i.e. at most one set-up change per period), needs to represent the following set of implications for δ^S_{rpt}:

$\delta^P_{rpt-1}\backslash\delta^P_{rpt}$	0	1
0	0	1
1	0	μ_{rpt}

$$\tag{3.5}$$

with

$$\mu_{rpt} := \begin{cases} 1, & \text{if any other production } p' \neq p \text{ is started in period } t-1 \\ 0, & \text{if no other production started in period } t-1 \end{cases}$$

(3.6)

These rules, for $\delta^P_{rpk} + \delta^P_{rpk-1} \neq 2$ are enforced by

$$\delta^S_{rpt} - \delta^P_{rpt}, \quad \forall \{rp\}, \quad t = 1$$

(3.7)

for the first period, and for all other periods $T_T := \{2, \ldots, T\}$ by

$$\delta^S_{rpt} \leq \delta^P_{rpt}, \quad \delta^S_{rpt} \geq \delta^P_{rpt} - \delta^P_{rpt-1}, \quad \forall \{rpt \in T_T\}$$

(3.8)

The case $\delta^P_{rpk-1} = \delta^P_{rpt} = 1$ is properly described by additional inequalities

$$\delta^S_{rpt} \geq -2 + \sum_{p' \neq p} \delta^S_{rp't-1} + \delta^P_{rpt} + \delta^P_{rpt-1}, \quad \forall \{rpt \in T_T\}$$

(3.9)

and

$$\delta^S_{rpt} \leq 2 + \sum_{p' \neq p} \delta^S_{rp't-1} - \delta^P_{rpt} - \delta^P_{rpt-1}, \quad \forall \{rpt \in T_T\}$$

(3.10)

If, in a general mode–product relation several products can be produced simultaneously, we may want that the case $\delta^P_{rpt-1} = \delta^P_{rpt} = 1$ also leads to $\delta^S_{rpt} = 0$; this can easily be realized by neglecting the second term on the right-hand sides of (3.9) and (3.10). Alternatively, we may require that $\mu = 1$ if any other more complicated rule than the above (3.7) is fulfilled.

Let us from now on assume that δ^P_{rpt} and δ^S_{rpt} are available. The production of product p may start in several time periods – i.e. we have several product-p-campaigns within the planning horizon T. Therefore we introduce continuous variables, $c_{rpt} \geq 0$, counting the number of start-ups and related to the start-up variables δ^S_{rpt} by

$$c_{rp1} = \delta^S_{rp1}, \quad \forall \{rp\}; \quad c_{rpt} = c_{rpt-1} + \delta^S_{rpt}, \quad \forall \{rpt \in T_T\}$$

(3.11)

Now we introduce continuous variables v_{rptn}, indicating whether a certain campaign, $n \in \mathbb{N}_0$, could be active ($c_{rpt} = n$) or not – i.e. whether

c_{rpt} is equal to a certain fixed integer $n \in \mathbb{N}_0$, or not. v_{rptn} represents the nonlinear function

$$v_{rptn} = 1 - \theta(|c_{rpt} - n|), \quad \theta(x) := \begin{cases} 1, & \text{if } x > 0 \\ 0, & \text{if } x = 0 \end{cases} \tag{3.12}$$

Let us assume that at most $N_{rp}^+ \in \mathbb{N}_0$ campaigns of product p can be produced within the planning horizon; a typical value in the current planning problem is $N_{rp}^+ = 6$. A special case is $N_{rp}^+ = 1$, enforcing that a product can be produced in only one campaign.

The relation (3.12) is enforced by

$$1 = \sum_{n=0}^{N_{rp}^+} v_{rptn} \quad \text{and} \quad \sum_{n=0}^{N_{rp}^+} n v_{rptn} = c_{rpt}, \quad \forall \{rpt\} \tag{3.13}$$

i.e. one campaign has to be chosen in any case (possibly, the '0' campaign), and if campaign n is selected then $c_{rpt} = n$. The sets

$$S_{rpt} := \{v_{rptn} | 0 \le n \le N_{rp}^+\}, \quad \forall \{rpt\} \tag{3.14}$$

form a special ordered set of type 1. We use the second equation of (3.13) as the reference row for efficient branching.

The total amount, p_{rpn}^C, of product p produced within campaign n is given by

$$p_{rpt}^C = \sum_{t=1}^{N^T} p_{rptn}^C, \quad \forall \{rp\}, \quad \forall n \in \mathcal{N}_1 := 1, \dots, N_{rp}^+ \tag{3.15}$$

where p_{rptn}^C is the amount of product p produced for campaign n in period t – i.e.

$$p_{rptn}^C = p_{rpt} v_{rptn}, \quad \forall \{rptn \in \mathcal{N}_1\} \tag{3.16}$$

Applying the formalism described in subsection 2.3 with $K = 1$ we replace (3.16) by

$$\begin{aligned} p_{rptn}^C &\le P_{rpt}^+ v_{rptn}, \quad p_{rptn}^C \le p_{rpt} \\ p_{rptn}^C &\ge p_{rpt} - P_{rpt}^+ + P_{rpt}^+ v_{rptn}, \quad \forall \{rptn\} \end{aligned} \tag{3.17}$$

With the formalism at hand described above we reached our goal: the computation of the amount, p_{rpn}^C, of product p produced for campaign n.

p_{rpn}^C may be now subject to specific batch constraints – e.g. a compaign may just consist of one single batch of fixed batch size B_{rp},

$$p_{rpn}^C = B_{rp}, \quad \forall \{rpn \in \mathcal{N}_1\} \tag{3.18}$$

Alternatively, campaigns may be built up by a discrete number of batches following each other immediately – i.e.

$$p_{rpn}^C = B_{rp}\beta_{rpn}, \quad \forall \{rpn \in \mathcal{N}_1\} \tag{3.19}$$

Where the integer variable β_{rpn} indicates the number of batches of size B_{rp} within campaign n. Finally p_{rpn}^C may behave like a semi-continuous variable – i.e.

$$p_{rpn}^C = 0 \quad \text{or} \quad C_{rp}^- \le p_{rpn}^C \le C_{rp}^+, \quad \forall \{rpn \in \mathcal{N}_1\} \tag{3.20}$$

where C_{rp}^- and C_{rp}^+ are lower and upper bounds if production takes place.

2.3. Modeling product terms including one continuous and several binary variables

To model products like $x\prod_{k=1}^K \delta_k$, where δ_k are binary variables and x is any kind of nonnegative variable, let us assume that X^+ is a valid upper bound on x. The product $\prod_{k=1}^K \delta_k$ is exactly represented by the variable y subject to the inequalities

$$\forall k: y \le X^+\delta_k, \quad y \le x, \quad y \ge x - X^+\left(K - \sum_{k=1}^K \delta_k\right) \tag{3.21}$$

The first inequality of (3.21) has the implications ($\delta_k = 0 \Rightarrow y = 0$) and ($y > 0 \Rightarrow \sum_{k=1}^K \delta_k = K$), while the second and third inequality give us ($\sum_{k=1}^K \delta_k = K \Rightarrow y = x$) and ($y = 0 \Rightarrow \sum_{k=1}^K \delta_k < K$). Note that if we want to know the product $y = x\prod_{k=1}^K \delta_k$ explicitly we do not need to introduce an extra variable.

2.4. Implementation and results

If we want to add the batch constraints in subsection 2.2 to the production planning model M1^2, it is not strictly necessary to use (3.3)–(3.10) to compute δ_{rpt}^P and δ_{rpt}^S. Alternatively, we can derive δ_{rpt}^P and δ_{rpt}^S from the mode state variables α_{rmt} and start-up variables β_{rmt} used in the model

M1 by Kallrath *et al.* (1994) and KW97 (320–4). If $\mathcal{P}$ is the union of disjunctive sets $\mathcal{P}_m$ of products produced in mode m and I_{rmp} indicates whether product p can be produced in mode m on reactor r, (in the current case we have $\Sigma_p I_{rmp} = 1$ – i.e. exactly one product per mode) we just have

$$\delta^P_{rpt} = \sum_{m|I_{rmp}=1} \alpha_{rmt}, \quad \delta^P_{rpt} = \sum_{m|I_{rmp}=1} \beta_{rmt}, \quad \forall \{rpt\} \tag{3.22}$$

This *special approach* based on (3.22) is, however, exactly identical only with the more *general approach* based on (3.7)–(3.10) if $P^-_{rpt} = 0$.

For the model M1 and a typical reference scenario (S_1) covering 12–36 production time periods we have used both approaches indicated by indices s and g to derive production plans maximizing total sales. The scenarios S_2 use (3.20) to model campaigns whose minimum size is 300 tons. The scenarios S_3 include 49 partial integer variables and use (3.19) to enforce that campaigns are built up by discrete batches of 100 tons each. Finally, in scenario S_m we require that if a certain mode is chosen the plant has to stay in that mode for at least three days. In this case, the variables m^D_{imk} used in KW97 (320–4), expressing how much time the plant at site i spends in mode m in period k, play the role of p_{rpt} used above; the length of the period (10–30 days) is a useful upper bound on m^D_{imk}. Using Dash's MILP-solver XPRESS-MP 10.05 (Ashford and Daniel, 1987, 1991), we got the following results (including the number of continuous, binary and semi-continuous variables, constraints, integer solution, number of nodes n_n, running timer τ, and gap Δ in percent) when we applied the formalism to all possible reactor (site)-product (mode)-time combinations:

	P^-_{prt}	n_c	b	$s-c$	c	IP	n_n	τ	Δ
S_1	−	12397	2973	1608	8441	1	440	8^m	1.9
S_1	−					2	960	$+6^m$	1.4
S_1	−					3	1721	$+8^m$	1.0
S_{2s}	−	14833	2973	1650	13997	1	786	52^m	19.4
S_{2g}	1	15217	2973	1650	14033	1	858	59^m	28.8
S_{2g}	$0.8P^+_{rpt}$	15217	2973	1650	14033	2	3860	5^h59^m	31.9
S_{3s}	−	15687	2973	1608	13855	1	632	58^m	3.9
S_{3g}	1	16215	2973	1608	15681	1	907	1^h09^m	16.5
S_{3g}	$0.8P^+_{rpt}$	16215	2973	1608	15681	3	18272	39^h02^m	4.6
S_m	−	15333	2973	1650	15681	1	511	28^m	4.6
S_m	−					3	1943	$+40^m$	1.8
S_m	−					4	3972	$+2^h44^m$	1.5

The use of special ordered sets of type 1 for the variables ν_{rptn} is essential; the model contains 192 sets and 1080 set members. In previous versions when these variables were declared as binary variables computing times were much larger. In the S_3 runs (multiple batches), the variables β_{rpn} were declared as partial integers (integer below 10, continuous above 10). Although the variables δ_{prt}^S become binary automatically, it is advantageous to declare them as binary variable explicitly because that enables us to prioritize them and to improve branching. The use of directives in the model was crucial. In the general approach scenarios n_n, τ, and the quality of the solution indicated by Δ depended critically on P_{rpt}^-. Note that the run for $P_{rpt}^- = 0.8\,P_{rpt}^+$, (i.e. high utilization rates of the plant system), shows that the third integer solution found and required much more computing time.

The benefit achieved by the extended model features is qualitative because it leads to an improved representation of the real-world process. The production plans do not suffer any longer from the time-indexed formulation and look more stable avoiding small campaigns and many set-up changes. In practical planning runs it is sufficient to use the formalism only for a few products or modes, and sometimes only for one site or reactor. Thus, the Pentium 166 MHz computing time reduces to less than 15 minutes and become similar to the one of the reference scenario S_1.

III. PRODUCTION PLANNING IN A PETROCHEMICAL PRODUCTION NETWORK

This section describes the mathematical model of a petrochemical production network. The network includes 5 plants [2 steamcrackers, 3 units to extract certain fractions] with about 30 products (or streams) in Ludwigshafen, and 7 plants with about 30 products in Antwerp. The crackers can be operated in continuously varying modes defined by the cracking severity. The modes of operation are modeled through interpolation between three predefined values (sharp, medium, mild). For a fixed mode of operation incoming and outgoing flows are linearly coupled by yield coefficients ($[t/t]$). Plants are subject to capacity restrictions (upper and lower bounds in $[t/h]$); for some plants there also exist lower or upper bounds for the relative weight of products in blends (cuts). In every time period only limited amounts of raw material are available. Streams from intern or extern sources are treated similarly. Utilities are also treated as products.

Every raw material or finished product can be transferred between sites; utilities, representing different types of energy (i.e. heating gas, electricity), cannot. It is possible to define different modes of transportation with different prices for every material. There exist minimal transport limits, below which transport is unreasonable. The solver can choose whether to transport nothing or an amount between the minimal limit, M_p^{TU} (t/h), and maximal transport capacity, M_p^{TL} (t/h).

The objective function is to maximize the net operating margin. Model output is the full feed stock information for all plants, the net operating margin for both networks, and shadow prices. The multi-period model can be used for monthly and six-months planning.

3.1. Foundations of the mathematical model

Indices, index sets, sets, and projectors

We use the following indices and set of indices:

c	$\in \{1, 2, 3\}$	$= C \subset \mathcal{N}$	crackers	$\lvert C \rvert$	$= 3$
d	$\in \{d_1, \ldots, d_{N^D}\}$	$= \mathcal{D}$	markets	$\lvert \mathcal{D} \rvert$	~ 5
m	$\in \{m_1, \ldots, m_{N^M}\}$	$= \mathcal{M}$	cracker modes	$\lvert \mathcal{M} \rvert$	$= 3$
n	$\in \{n_1, \ldots, n_{N^N}\}$	$= \mathcal{N}$	nodes	$\lvert \mathcal{N} \rvert$	~ 20
o	$\in \{o_1, \ldots, o_{N^O}\}$	$= O$	processes	$\lvert O \rvert$	~ 80
p	$\in \{p_1, \ldots, p_{N^P}\}$	$= \mathcal{P}$	products	$\lvert \mathcal{P} \rvert$	~ 40
r	$\in \{r_1, \ldots, r_{N^R}\}$	$= \mathcal{R}$	resources	$\lvert \mathcal{R} \rvert$	~ 30
s	$\in \{s_1, \ldots, s_{N^S}\}$	$= S$	sites	$\lvert S \rvert$	$= 2$
t	$\in \{1, \ldots, N^T\}$	$= \mathcal{T}$	time periods	$\lvert \mathcal{T} \rvert$	$= 6$

Processes – e.g. cracking of LPG in a certain mode of operation or production of Benzol through extraction from the blend BTX – are represented by nodes (called *sub-models* in the PIMS[3] nomenclature). Because additional equations are defined in the crackers, special indices are defined for them. We use the term *products* for all streams considered in the planning model.

Different modes of operation are possible only in the crackers. Therefore the indices for the modes are not used and instead three different processes are defined. They correspond to the three cracker modes mentioned above. Resources stand for capacity restrictions. It is possible to share resources among processes. Usually resources are identical to certain minimal or maximal node flow constraints. We do not introduce an additional index for transport because we use only one.

For simplification of description we introduce the following sets:

$\mathcal{B}$ blends
$\mathcal{L}$ stored streams
$\mathcal{O}_{mc}$ processes belonging to mode m in cracker c
$\mathcal{P}_F$ finished products
$\mathcal{P}_T$ products transported between sites
$\mathcal{P}_R$ raw materials purchased
$\mathcal{P}_P$ finished products purchased

Additionally we define the following projectors:

$\mathcal{P}^P_{P_b}$ the blended products $p \in \mathcal{P}$ for every blend (cut) $p_b \in \mathcal{B}$
$\mathcal{P}^O_n$ the corresponding processes $o \in \mathcal{O}$ for every node $n \in \mathcal{N}$
$\mathcal{P}^O_b$ the corresponding processes $o \in \mathcal{O}$ for $b \in \mathcal{B}$
$\mathcal{P}^N_b$ the corresponding nodes $n \in \mathcal{N}$ for $b \in \mathcal{B}$
$\mathcal{P}^O_s$ the processes at site $s \in S$

Finally, we define the length of periods. For every period its length is given by F_t (h/month). This increases model flexibility, because it provides an easy way to change from monthly to weekly planning. Additionally, different lengths of months can be incorporated.

Variables

We use the following set of variables:

b_{pp_bts} Blending of product p [t/h] into product p_b in period t
e_{pts} Amount purchased [t/h] of product p at site s in period t
f_{ot} Flow [t/h] of process o in period t; process flow variable
i_{pts} Stock [t] of product p ($p \in \mathcal{L}$) at the end of period t
q^C_{pts} Collector for pool p in period t at site s
$q^D_{p_bnts}$ Flow of pool p_b into node n in period t
r_{pp_bts} Recursion error for the amount of product p in product p_b
s^j_{its} Variable for interpolation of severity ($i, j \in \{1, 2, 3\}$)
$t_{ps_is_jt}$ Transport of product p from site s_i to site s_j in period t
v_{ptsd} Sold [t/h] product p at site s in period t on market d
$\lambda_{ps_is_jt}$ Binary variable for transport of product p from s_i to s_j

3.2. The mathematical model

The model consists of a network of nodes connected by appropriate balance equations defining and representing the topology. The *flow balances*

$$e_{pts} + t_{ps_js_jt} - t_{ps_is_jt} + \sum_{o \in \mathcal{P}^O_{s_i}} Y_{op} f_{ot} = \sum_d v_{ptsd} \lesseqgtr 0, \quad \forall \{pts_i\} \tag{3.23}$$

typically include the sum of purchases, transport, consumption and production, and sells; inventories will be considered later. Consumption and production are represented by positive und negative values of yield coefficients Y_{op} coupled to the process flow variables f_{ot}. Y_{op} with o identifying a process defines how much of product p is consumed ($Y_{op} > 0$) or generated ($Y_{op} < 0$). Necessarily $\Sigma_p Y_{op} = 0$. For convenience, yield coefficients are normalized to $\Sigma_p |Y_{op}| = 2$, so that the sum of the positive and negative entries is 1.

A similar balance relation must hold for the utilities. Both crackers use heating gas, in this case a mixture of CH_4 and H_2, and electricity to heat up some feed stock products. Consumption of utility p in process o is given by H_{op} which leads us to the relations

$$e_{pts} + \sum_{o \in \mathcal{P}^O_s} H_{op} f_{ot} \lesseqgtr 0, \quad \forall \{pts\} \tag{3.24}$$

between the purchase variables, e_{pts}, and the process flow variables, f_{ot}. For numerical advantages the equalities (3.24) are relaxed and equivalently replaced by inequalities (for further reasoning and validation see KW97, 281).

For some products p, inventories with initial stock S^A_{ps} have to be considered – i.e., at the end of the first period the inventory is the sum of the opening inventory and the incoming flow minus the outgoing flow:

$$i_{pts} = S^A_{ps} + F_t f^N_{pts}, \quad \forall p \in \mathcal{L}, \quad t = 1, \quad \forall s_i \tag{3.25}$$

with the number, F_t, of hours in period t and

$$f^N_{pts} := e_{pts} + t_{ps_js_i} - t_{ps_is_j} + \sum_{o \in \mathcal{P}^O_{s_i}} Y_{op} f_{ot} = \sum_d v_{ptsd} \tag{3.26}$$

In the inventory balance equations applied to the end of the periods $t \in \mathcal{T}_2 = \{2, \ldots, N^T - 1\}$

$$i_{pts} = i_{pt-1m} + F_t f^N_{pts}, \quad \forall p \in \mathcal{L}, \quad t = \mathcal{T}_2, \quad \forall s_i \tag{3.27}$$

the initial inventory is replaced by the inventory of the previous period. In the last period the target inventory S_p^E (at present, the implementation assumes $S_{ps}^A = S_{ps}^E = 0$) is used, so we get

$$i_{pts} = i_{pt-1m} + F_t f_{pts}^N = S_{ps}^E, \quad \forall p \in \mathcal{L}, \quad t = N^T, \quad \forall s_i \tag{3.28}$$

Modelling of blends and pooling: Blends (blended products), or cuts are flows, which consist of several components, but behave as one stream topologically. A cut is characterized by its components and its composition which is variable to a certain amount. Cuts are treated as products and their components as the cut's properties. Cuts can be components of other cuts. Two different types of cuts are used, described by their own set of variables, although mathematically they are similar. The first type – products leaving the sub-models – are in fact blends and separated into their components later, the second type are products pooled in blending nodes.

Equations for sub-model nodes need to connect at first the pool collector variable $q_{p_b ts}^C$ (this is the total mass of the produced blend) to the yield coefficients and flow variables

$$q_{p_b ts}^C = \sum_{p \in \mathcal{P}_{p_b}^P} \sum_{o \in \mathcal{P}_s^O} Y_{op} f_{ot}, \quad \forall p_b \in \mathcal{B}, \quad \forall \{ts\} \tag{3.29}$$

For every component the error vectors $r_{pp_b ts}$ denote the deviation of the total amount, $\Sigma_{o \in \mathcal{P}_s^O} Y_{op} f_{ot}$, of the component p in the blend from the value computed as the product of $q_{p_b ts}^C$ and the assumed value, $F_{pp_b ts}^C$, for its relative fraction. Thus we get [compare (3.30) to (11.1.21) in KW97, 371]

$$t_{pp_b ts}^{CC} := F_{pp_b ts}^C q_{p_b ts}^C + r_{pp_b ts} = \sum_{o \in \mathcal{P}_s^O} Y_{op} f_{ot}, \quad \begin{matrix} \forall p_b \in \mathcal{B} \\ \forall p \in \mathcal{P}_{p_b}^P \\ \forall \{ts\} \end{matrix} \tag{3.30}$$

The quantity $t_{pp_b ts}^{CC}$ describes the linear approximation of the amount of p in the blend p_b. The error vector is used in subsequent processes involving the blend. The guessed quantity $F_{p_b nt}^D$ in (3.31) specifies which fraction of the error vector flow is distributed to the node belonging to the processes:

$$F_{pp_b ts}^C q_{p_b nt}^D + F_{p_b nt}^D r_{pp_b ts} = \sum_{o \in \mathcal{P}_n^O} Y_{op} f_{ot}, \quad \forall \{st\}, \quad \begin{matrix} \forall n \in \mathcal{P}_s^N \\ \forall p_b \in \mathcal{B} \\ \forall p \in \mathcal{P}_{p_b}^P \end{matrix} \tag{3.31}$$

Equations for the blending nodes again connect the pool collector variable $q^C_{p_b ts}$ with the flow-in variables. For blends consisting only of base products, and for blends containing other blends we have two different formulae, namely

$$\sum_{p \in \mathcal{P}^P_{p_b}} b_{pp_b ts} = q^C_{p_b ts} \quad \text{and} \quad \sum_{p \in \mathcal{P}^P_{p_b}} F^C_{pp_b ts} b_{pp_b ts} + F^D_{p_b nt} r_{pp_b ts} = q^C_{p_b ts}$$

$$\forall p_b \in B, \quad \forall \{ts\} \tag{3.32}$$

The equations for the error vectors $r_{pp_b ts}$ are:

$$b_{pp_b ts} = t^{CC}_{pp_b ts} \quad \text{and} \quad F^C_{pp_b ts} b_{pp_b ts} + F^D_{p_b nt} r_{pp_b ts} = t^{CC}_{pp_b ts}$$

$$\forall p_b \in \mathcal{B}, \quad \forall p \in \mathcal{P}^P_{p_b}, \quad \forall \{ts\} \tag{3.33}$$

In the case of a blend flowing into a sub-model we again have again (3.31).

In *distributive recursion*, an equivalent technique to sequential linear programming, after every solution of the linear problem the guesses for concentrations and distributions are exchanged and a new iteration $(k + 1)$ is started, until convergence is achieved:

$$F^{C,(k+1)}_{pp_b ts} = F^{C,(k)}_{pp_b ts} + r^{(k)}_{pp_b ts} \Big/ q^{C,(k)}_{p_b ts}, \quad \begin{array}{c} \forall p_b \in \mathcal{B} \\ \forall p \in \mathcal{P}^S_{p_b} \\ \forall \{ts\} \end{array} \tag{3.34}$$

$$F^{D,(k+1)}_{pnt} = q^{D,(k)}_{p_b nt} \Big/ q^{C,(k)}_{p_b ts}, \quad \begin{array}{c} \forall p_b \in \mathcal{B} \\ \forall \{ts\} \\ \forall n \in \mathcal{P}^N_s \end{array} \tag{3.35}$$

Initial guesses $F^C_{pp_b ts}$ for concentrations and $F^D_{p_b nt}$ for distributions are used to start the iteration.

Concerning *blends (cuts)* we need to restrict the composition of some cuts. The numbers $F^{EL}_{pp_b t}$ und $F^{EU}_{pp_b t}$ represent lower and upper bounds on the relative fractions of components. The quantities $F^{C,(k+1)}_{pp_b s}$ and $F^{D,(k+1)}_{pnt}$ must observe these bounds.

Capacity restrictions apply to all processes consuming R_{or} (units/h) of available capacity resources. Additionally lower and upper bounds D^{LR}_r and D^{UR}_r on the total capacity consumption are needed. The sum of all processes using the same resource is then limited by:

$$D_r^{LR} \leq \sum_o R_{or} f_{ot} \leq D_r^{UR}, \quad \forall r \tag{3.36}$$

To model the *interpolation of cracker modes*, cracking severity variables s_{cmt} for every operating mode of the crackers are introduced ($0 \leq s_{cmt} \leq 1$). They describe the percentage at which the cracker works in the corresponding mode (sharp or medium or mild). The process variables, f_{ot}, are coupled to the severity variables by

$$\sum_{o \in O_{mc}} f_{ot} \leq R_c s_{cmt}, \quad \forall \{cmt\}; \quad \sum_{m=1}^{3} s_{cmt} = 1, \quad \forall \{ct\} \tag{3.37}$$

If the resource R_c is the total capacity of the cracker, and if the crackers operate at full capacity the interpolation is exact if we proceed as follows: in the case of only three modes the cracking severity variables are connected to binary variables

$$\mu_{ct} := \begin{cases} 1, & \text{in period } t \text{ cracker } c \text{ operates} \\ & \text{between mode 1 and 2,} \quad \forall c \in \{1,2\} \\ 0, & \ldots \text{between mode 2 and 3} \end{cases} \tag{3.38}$$

i.e., we interpolate only between two neighboring modes (sharp, medium or medium, mild):

$$s_{c1t} \leq \mu_{ct}, \quad s_{c3t} \leq 1 - \mu_{ct}, \quad \forall \{ct\} \tag{3.39}$$

This formulation enables us to compute the cracking severity ε_{ct} by

$$\varepsilon_{ct} = 0.5 s_{c1t} + 0.55 s_{c2t} + 0.6 s_{c3t}, \quad \forall \{ct\} \tag{3.40}$$

where the reference values 0.5, 0.55 und 0.6 represent the three modes.

Bounds

The *cracking severities* ε_{ct} are bounded by S_c^{LS} and S_c^{US}. *Initial inventories*, S_{ps}^{A}, (t) for the first period in and target closing inventories, S_{ps}^{E}, (t) for the last period are assumed to be zero. *Inventory capacities* (t) are defined by S_{ps}^{C} and have to be observed by the inventory variables i_{pts}.

Availability Restrictions and Purchase All materials are bought on the same market, but at possibly different prices and bounds for the sites.

The purchased streams e_{pts} are subject to availability restrictions – i.e., to upper and lower bounds (t/h) E_{pts}^U and E_{pts}^L. Sales restrictions put bounds (t/h) V_{ptsd}^U and V_{ptsd}^L on the sold streams v_{ptsd}. Lower bounds represent given contracts with demand to be satisfied. There is the possibility to fulfill these obligations by buying products from extern sources instead of producing them. This type of purchase is treated separately.

Transport restrictions consider that a certain minimal amount has to be transported, if transport is to take place at all:

$$M_p^{TL} \lambda_{ps_is_jt} \le t_{ps_is_jt} \le M_p^{TU} \lambda_{ps_is_jt}, \quad \forall \{pts_i\}, \quad \forall s_j \neq s_i \tag{3.41}$$

Unfortunately it is not possible to implement (3.41) into PIMS, because transport variables cannot (at present) be connected to other variables. Therefore we have to define the restrictions in the local models, for example

$$M_p^{TL} \lambda_{ps_is_jt} \le b_{pp_bts_j} \le M_p^{TU} \lambda_{ps_is_jt}, \quad \forall \{pts_i\}, \quad \forall s_j \neq s_i \tag{3.42}$$

for a product absorbed to 100% by a blend (otherwise the corresponding sum has to be used in the left-hand side), or

$$M_p^{TL} \lambda_{ps_is_jt} \le f_{ot} \le M_p^{TU} \lambda_{ps_is_jt}, \quad \forall \{pts_i\}, \quad \forall s_j \neq s_i \tag{3.43}$$

for a product absorbed to 100% by a certain process.

Objective function

The objective function includes terms for revenue, product consumption, external purchase, transport, and inventory. Revenues and purchases have to be multiplied with the factor, F_t, because the corresponding data is defined on an hourly basis. The costs for holding inventory is defined on a monthly basis. F_t^{PV} describes the effects of inflation and interest rates in the model. As a side-effect, some symmetries in the model are broken, which leads to better convergence. This effect is added to by the usage of inventory holding costs. The objective function is formulated as

$$z = \sum_{t \in T} F_t^{PV} [F_t(y_t^E - y_t^R - y_t^P - y_t^T) - y_t^L] \tag{3.44}$$

total revenue z_t^E based on specific revenues E_{ptsd}^P (*DM/t*),

$$z_t^E := \sum_{p \in \mathcal{P}_F} \sum_{s \in S} \sum_{d \in D} E_{ptsd}^E v_{ptsd} \tag{3.45}$$

total cost costs for raw materials, z_t^R, and products purchased externally

$$z_t^R := \sum_{p \in \mathcal{P}_R} \sum_{s \in S} C_{pts}^R e_{pts}, \quad z_t^P := \sum_{p \in \mathcal{P}_T} \sum_{s \in S} C_{pts}^P e_{pts} \tag{3.46}$$

based on specific costs E_{ptsd}^R and C_{pts}^E (*DM/t*) for materials consumed and purchased externally, and transport and inventory costs

$$z_t^T := \sum_{p \in \mathcal{P}_P} \sum_{s_i \in S} \sum_{\substack{s_j \in S \\ s_j > s_i}} C_{pzs_is_j}^T t_{pzs_is_j}, \quad z_t^L := \sum_{p \in \mathcal{L}} \sum_{s \in S} C_{pts}^L i_{pts} \tag{3.47}$$

The specific transport costs are given for every material specifically as $C_{pzs_is_j}^T$ (*DM/t*). The specific costs for inventories C_{pts}^L (*DM/t*) are calculated from the cost of the working capital tied up in the inventory. Prices for sold and purchased products are given on a monthly base (*DM/t*). They can be either 'official' market prices or intern 'computational' prices.

3.3. Solution approach and results

The model falls into the class of MINLP problems. The mathematical algorithm implemented in PIMS is similar to a first step of outer approximation (see subsection 4.2). At first, the NLP relaxation is solved by *distributive recursion* which is equivalent to sequential linear programming but has better scaling (see, for instance, KW97, 368). The next step is to fix the recursed terms (concentrations) and to solve the MILP problem. Finally, with the discrete variables fixed, an NLP problem is solved updating the recursed terms.

The matrix of the resulting linear programs (only Ludwigshafen) has about 1300 rows, 1600 columns and 15 500 nonzeros. 30 of the variables are binary, and we have 269 nonlinear constraints. Antwerp adds another 800 rows and 1000 columns. Solution times are between 2 and 15 minutes. Once a solution has been found the recursed terms are stored and used as initial values in subsequent runs.

Typical questions analyzed are for example: 'which product should be produced at which site?', 'which raw materials should be purchased?', 'under which circumstances is it advantageous to transfer products between the sites?', or 'which effects do certain changes have on the global system?'. Several case studies have been performed for the individual sites and for the two-site-network. The savings are generally in the order of magnitude of 1%, which corresponds to several million DM per year.

IV. AN INTEGRATED SITE ANALYSIS

The purpose of the model is to design an integrated production network minimizing the costs for raw material (RM), investment and variable costs for reprocessing units, and a cost penalty term for remaining impurities.

Three types of production processes (units) are considered: source processes producing RM not requesting it, sink processes only requiring RM not producing it, and stream processes requesting and producing the RM. The flow rate of a process and its effect on the quality of RMs are known *a priori*.

Until now, purchased RM of different qualities measured in terms of certain impurities has been used for all production processes requiring RM. It seems recommended to reuse impure RM for other processes and so to reduce the costs for purchasing RM. This may require new connections between units. Unfortunately, for most processes inlet specifications for the RM restrict the direct reuse of impure RM of other processes. In addition, some flow connections between processes are not allowed or not possible ('forbidden matches'). However, the inlet specifications may be satisfied by pooling RM streams of different quality. This has the additional advantage that only one pipeline is necessary instead of several so that investment costs for pipelines (depending on required capacity and the distance between the production processes to be connected) and costs for pumping RM into pipelines are reduced. For each single process we know the amount and type of impurities it produces – i.e., has to be added to the impurities already in the stream. To pool the RM streams no *investment or variable costs* have to be taken into account, except for the case of new pipelines to be built because pools are realized by joining different RM streams without any technical or financial expense.

An alternative approach is to process the quality of the RM when leaving a process. This requires local *reprocessing units* (RPUs) for a (partial) improving of the RM quality, causing investment costs for RPUs, as well as costs for operating the RPUs. The components extracted by the RPUs might be reused or sold. Since for certain impurities we consider penalty costs when they leave the system money can be saved by reducing the amount of these impurities.

A central RPU already exists which improves the quality of RM before leaving the site; the RM is used by other sites but the more impurities remain leaving the site the lower the quality, and thus the less valuable it is. Since we cannot easily convert quality into money we consider penalty costs for the impurities remaining in the RM. Therefore, it might be more promising building small local RPUs.

The investment costs for new local RPUs depend on the required capacities (RM flow rates) but additionally on the type of impurity and on its concentration. The variable costs for the RPUs depend on the mass load (in kg/h) of the input impurities. For the use of the central RPU only variable costs have to be paid, depending on the type and the total mass of the impurity.

The RPUs can be regarded as a certain type of stream process, where the total input and output of RM and impurities are the same. Unlike production stream processes RPUs split both RM stream and mass load of impurities – i.e., two RM streams leave the RPU: a main RM stream (relatively high quality) and a small stream (very low quality). The ratio of the amount of RM of the main outlet stream to that of the input stream is prescribed for each RPU. Analogously the ratio ('extraction rate') of the mass load of an impurity in the low-quality stream to those of the inlet mass load can be estimated as a constant or prescribed as a function of the inlet concentration for each RPU and each impurity. Similar to production processes, for technical reasons, there might exist inlet specifications limiting the concentration of a certain impurity. Since the raw material leaving the production network might be used for other purposes we also consider bounds for the concentration of impurities in the RM leaving the system. They can be understood as 'outlet specification' (i.e., maximum concentration of an impurity in the RM) of the central RPU.

4.1. The mathematical model

At first let us summarize the dimensions of the model and the indices:

	#	description		
			i	sources
	3	raw material qualities	s	sinks
$N \sim 60$		source processes	k	impurities
$M \sim 7$		impurities	m	reprocessing units
$L \geq 1$		RPUs	p	pools
$P \sim 60$		pools	s	connection capacity

$$(3.48)$$

We consider a total of $3 + N + L + P \sim 125$ processes, and introduce, for convenience, the following sets of indices

$$
\begin{array}{ll}
\mathcal{K}^P & \text{set of all pairs } (i, j) \text{ of possible matches} \\
\mathcal{K}^N & \text{set of all pairs } (i, j) \text{ of unexisting matches } (\mathcal{K}^N \subset \mathcal{K}^P) \\
\mathcal{P}^{SO} & i \in \mathcal{P}^{SO} \Leftrightarrow P_i \text{ is a source process} \\
\mathcal{P}^{SI} & j \in \mathcal{P}^{SI} \Leftrightarrow P_j \text{ is a sink process} \\
\mathcal{P}^{ST} & i \in \mathcal{P}^{ST} \Leftrightarrow P_i \text{ is a stream process} \\
\mathcal{P}^T & m \in \mathcal{P}^T \Leftrightarrow P_m \text{ is a reprocessing unit} \\
\mathcal{P}^P & m \in \mathcal{P}^P \Leftrightarrow P_p \text{ is a pool} \\
\mathcal{P}_1 & \mathcal{P}_1 := \mathcal{P}^{SO} \cup \mathcal{P}^{ST} \cup \mathcal{P}^P \cup \mathcal{P}^T \\
\mathcal{P}_2 & \mathcal{P}_1 := \mathcal{P}^{SI} \cup \mathcal{P}^{ST} \cup \mathcal{P}^P \cup \mathcal{P}^T
\end{array} \tag{3.49}
$$

Variables

real variable	dim.	description
$x_{ij}^A \equiv x_{ij}$	t/h	output from process P_i going to process P_j (at waste RM RPUs: output (A), high quality)
$x_m^B = x_{im}^A(1 - G_m)$	t/h	low-quality output (B) from process P_i; for other processes than local RPUs $x_i^B = 0$
$c_{ik}^{in}(c_{ik}^{out})$:	ppm	input (output) concentration of impurity k in process P_i
$z_{jk}^{in} := \sum_{\substack{i \\ (i,j) \in K^P}} C_{ik}^{out} x_{ij}$	t/h	input mass load of impurity k in process P_j summed over all processes P_i which send RM to it

In addition we use the binary variables μ_{ij}, ε_{sij}, and ν_m indicating whether a connection exists from P_i to P_j, whether a pipeline from P_i to P_j is of capacity and whether process (RPU) P_m exists.

data	dim.	description
K_{ik}^{out}	[ppm]	specific outlet concentration of impurity k of process P_i
K_{jk}^{in}	[ppm]	inlet specification for impurity k of process P_j
X_i	[t/h]	inlet flux of process P_i, $X_i = 0$ for source processes P_i

data	dim.	description
Y_i	[t/h]	outlet flux of process P_i, $Y_i = X_i$ for all stream processes and $Y_i = 0$ for sink processes; for local RPUs it is $Y_i \leq X_i$ because the RM stream is splitted into a main stream (A) and a smaller low-quality stream (B)
$F_{mk}(c_{mk}^{in})$	[–]	rate of extraction for a single impurity k removed in RPU P_m. This rate is a function of the input concentration c_{mk}^{in} and is of order 0.7 . . . 0.95
G_m	[–]	splitting rate of the RM stream within RPU m
C_{ij}^{PI}		investment costs for building a pipeline from P_i to P_j
$C_m^{TI}(x_{im}, z_{mk}^{in})$		investment costs for a local RPU P_m; these costs are a function of the RM flow rate once it has been decided which impurities have to be removed
$C_{mk}^{TV}(z_{mk}^{in})$		variable costs for RPU P_m, they are a function of the product of extraction rate and mass load of impurity k – i.e., $f_{mk}(c_{mk}^{in})z_{mk}$
C_i^{RM}		costs for purchased RM of quality i
C_k^{PEN}		penalty costs for impurities leaving the system
V_s^{PI}	[t/h]	capacity of pipeline is of type (size) s

Constraints

The production processes (sink and stream processes) require a constant amount of RM, and the production (source or stream) processes have a constant emission of waste RM. Therefore the sum of all inlet and outlet fluxes is constant, and we have the *mass balances*

$$\sum_{i|(i,j)\in K^P} x_{ij} = X_j, \quad \forall j \in \{\mathcal{P}^{SI} \cup \mathcal{P}^{ST}\} \tag{3.50}$$

and

$$\sum_{j|(i,j)\in K^P} x_{ij} = Y_i, \quad \forall i \in \{\mathcal{P}^{SO} \cup \mathcal{P}^{ST}\} \tag{3.51}$$

Additionally, for stream and pool processes it is assumed that there is no loss of RM – i.e.,

$$x_i = Y_i, \quad \forall i \in \{\mathcal{P}^P \cup \mathcal{P}^{ST}\} \tag{3.52}$$

In RPUs the RM stream is separated into a main stream (A) of relatively high quality and a stream (B) of low quality. Since no RM loss is assumed the RM flow of the RPU P_m can be modeled similar to those of stream processes:

$$\sum_{i|(i,m)\in K^P} x_{im} = \sum_{j'|(m,j')\in K^P} x_{m,j'} + x_m^B, \quad \begin{array}{l} \forall i \in \{\mathcal{P}^{SO} \cup \mathcal{P}^{ST} \cup \mathcal{P}^P\} \\ \forall j \in \{\mathcal{P}^{SI} \cup \mathcal{P}^{ST} \cup \mathcal{P}^P\} \\ \forall m \in P^T \end{array} \tag{3.53}$$

The splitting of the main RM stream into two streams within a RPU plant is prescribed by the factor G_m, the ratio of RM flow rate in stream (B) to the total input flow rate ($G_m = 1$ implies that all RM is kept in the main stream):

$$G_m = \frac{x_m^B}{\sum_{i|(i,m)\in K^P} x_{im}} \Leftrightarrow x_m^B = \sum_{i|(i,m)\in K^P} G_m x_{im}, \quad \forall m \in \mathcal{P}^T \tag{3.54}$$

The pool P_j collecting different RM streams originating from processes P_i is described by

$$c_{jk}^{in} x_j = \sum_{i|(i,j)\in K^P} c_{ik}^{out} x_{ij}, \quad \begin{array}{l} \forall i \in \mathcal{P}_1 \\ \forall k, \forall j \in \mathcal{P}^P \end{array} \tag{3.55}$$

The concentration limits for certain impurities originating from the processes P_i and entering P_j yield the material balance with respect to the content of impurities:

$$\sum_{i|(i,j)\in K^P} c_{ik}^{out} x_{ij} =: z_{jk}^{in} \le K_{jk}^{in} X_j, \quad \forall j \in \mathcal{P}_2, \quad \forall k \tag{3.56}$$

Knowing the concentration of impurity k and the RM fluxes originating from process P_i and entering process P_j the pooled concentration c_{jk}^{in} – i.e., the inlet concentration of P_j – can be calculated by

$$c_{jk}^{in} \sum_{i|(i,j)\in K^P} x_{ij} = \sum_{i|(i,j)\in K^P} e_{ik}^{out} x_{ij} \tag{3.57}$$

$$\Leftrightarrow c_{jk}^{in} X_j = z_{jk}^{in}, \quad \forall j \in \mathcal{P}_2, \quad \forall k \tag{3.58}$$

The outlet mass load of impurities of a process consists of the impurities which already have been in the inlet RM stream and the impurities added by the production process. Their total concentration in the outlet stream can be calculated by

$$c_{jk}^{out} = c_{jk}^{in} + K_{ik}^{out}, \quad \forall j \in \{\mathcal{P}^{SO} \cup \mathcal{P}^{ST}\}, \quad \forall k \tag{3.59}$$

and the output mass load of stream processes (not valid for RPUs!) is

$$z_{ik}^{out} X_i = z_{ik}^{in} + K_{ik}^{out} X_i \tag{3.60}$$

The outlet concentration of a impurity k of a RPU P_j depends on the inlet concentration and the extraction rate for removing the substance. As already mentioned above the impurities can be divided in two groups with additional sub-groups for which the removal in RPUs is very different. In general it can be described by the function $z_{jk}^{out} = H(z_{jk}^{in})$: The left and right formulae show typical relations representing the operation of the RPUs:

$$z_{jk}^{out} = z_{jk}^{in}\left(1 - F_{jk}\left(z_{jk}^{in}\right)\right) \qquad z_{jk_1}^{out} = z_{jk*}^{in} - z_{j,k_2}^{in}$$

$$\text{or} \qquad\qquad\qquad z_{j,k_2}^{out} = 0$$

$$z_{j1}^{out} = z_{j1}^{in} - \sum_{k=2}^{6} z_{jk}^{in} F_{jk}\left(z_{jk}^{in}\right) \quad z_{j,k_2}^{out} = z_{j,k_3}^{in} \tag{3.61}$$

The construction of pipelines from i to j can be required by the inequalities

$$x_{ij} \leq \min_{ij}\{Y_i, X_j\}\mu_{ij}, \quad x_{ij} \leq \sum_{s} V_{sij}^{PI} \varepsilon_{sij}, \quad \forall (i,j) \in \mathcal{K}^N \tag{3.62}$$

where the second inequality describes the required type (capacity) of pipeline.

To be sure that only one pipeline is built between process P_i and P_j the sum over s over the binary variable ε_{sij} which describes the type (capacity) of pipeline is forced to be μ_{ij} which is 1 if the connection exists and which is 0 otherwise:

$$u_{ij} = \sum_{s=1}^{S} \varepsilon_{sij}, \quad \forall i \in \mathcal{P}_1, \quad \forall j \in \mathcal{P}_2 \tag{3.63}$$

In order to force the construction of a RPU P_m the inequalities

$$\sum_{i|(i,m)\in K^P} x_{im} \le \left(\sum_{j'|(m,j')\in K^P} x_{m,j'} + x_m^P \right) v_m, \quad \begin{array}{l} \forall i \in \mathcal{P}^{SO} \setminus \mathcal{P}^T \\ \forall m \in \mathcal{P}^T \\ \forall t \end{array} \tag{3.64}$$

have to be fulfilled.

Objective function

The objective function sums over all investment and variable costs, over penalty costs to be paid and over the income which can be achieved by reusing impurities and it considers all possible matches between the processes. This sum is to be minimized:

$$\begin{aligned} Z := \min \sum_{\substack{j \quad i \\ (i,j)\in K^P}} & \sum C_{ij}^{PI} \mu_{ij} \\ & + \sum_{\substack{j \quad i \\ (i,j)\in K^P}} \sum \left\{ C_j^{TI}(x_{ij}, z_{jk}^{in}) v_j \sum_k C_j^{TV} z_{jk}^{in} \right\} \\ & + C_{ij}^{RM} x_{ij} - \sum_{\substack{j \quad k \\ (i,j)\in K^P}} \sum S_k (z_{jk}^{in} - z_{jk}^{out}) \\ & + \sum_k C_k^{PEN} z_{ctr,k}^{out} \end{aligned} \tag{3.65}$$

The first term in (3.65) represents the investment costs for pipelines from process i to j, the third the variable costs and investment costs for RPU j collecting streams from process i, and the fourth term the costs for RM streaming from process i to j, $i \in \{1, 2, 3\}$. The second-last term is the revenue from sold impurity k in streams originating from processes i and extracted in process j, the last term represents the costs for impurity k leaving the system after having passed the central RPU P_{ctr}.

4.2. Solution approach and results

Mathematical solution – outer approximation

To solve this MINLP problem we use the *outer approximation* (OA) algorithm by Duran and Grossmann (1986). This algorithm generates a sequence of NLP sub-problems (produced by fixing the binary variables y^k) and MILP Master problems. Algorithms based on OA describe the feasible region as the intersection of an infinite collection of sets with a simpler structure – e.g. polyhedra. In OA the Master problems are generated by 'outer approximations' (linearizations, or Taylor series expansions) of the nonlinear constraints at *those points* which are the optimal solutions of the NLP subproblems. The key idea of the algorithm by Duran and Grossmann (1986) is to solve the MINLP with a much smaller set of points – i.e. tangential planes. In convex MINLP problems, a super-set of the feasible region is established. Thus, the OA Master problems (MILP problem in both discrete and continuous variables) produce a sequence of lower bounds monotonically increasing. The NLP sub-problems yield upper bounds for the original problem while the MILP Master problems yield additional combination of binary variables y^k for subsequent NLP sub-problems. Under convexity assumptions the Master problems generate a sequence of lower bounds increasing monotonically. The algorithm terminates if lower and upper bounds equal or cross each other. The OA algorithm has heuristic extensions for nonconvex MINLP.

Software

To model and solve the MINLP the software package GAMS by GAMS Inc. (Washington) (see Brooke *et al.*, 1992) with the DICOPT-algorithm (Viswanathan and Grossmann, 1990) using a nonlinear solver (CONOPT by ARKI Consulting & Development A/S, Denmark) in combination with a MILP-solver are used. DICOPT (Viswanathan and Grossmann, 1990) seems to be the only commercial software available for solving the MINLP problem (1.1) of realistic size. It uses OA with some extensions for nonconvex problems. To initialize the algorithm the first linearization is derived from the solution of the continuous relaxation of the MINLP – i.e. it is not required that the user provides any discrete initial point. The termination criterion is different from a pure 'crossing bounds' method. In a nonconvex model the algorithm terminates when the solutions of the NLP problems do not provide improved upper bounds.

Homotopy method

The problem is solved in sequence of sub-models formulated in GAMS, exploiting the results of the previous one – i.e. we use a homotopy method. A simple linear model provides initial values for two simple nonlinear sub-models. Solution times are of the order of 1 or 2 hours.

Results

The model was well appreciated by the client for its high degree of reality, exact mass balances of raw material impurities, and the free pools. The model reproduced and confirmed earlier suggestions by engineers establishing a certain amount of trust, and finally, suggested further nonintuitive improvements with remarkable financial savings.

V. A PRODUCTION PLANNING AND PROCESS DESIGN PROBLEM

This optimization problem is concerned with a production process of a certain product P involving a system of connected reactors. The problem is a typical process design problem leading to a mixed-integer nonlinear model. Nonlinear terms are related to the exponential terms for the reaction kinetics and rational terms to describe the mass flow. The description of reaction kinetics is in part based on nonlinear expression (interpolated and approximated functions describing density and viscosity). Discrete features are needed to count the number of reactors, the existence of connections, the length of reactor chains, and to select the size of reactors. The variables are the flow rates, fractions, and the number and size of reactors.

5.1. Mathematical formulation of the model

Throughout this model description the following set $r \in \mathcal{R} \cup \mathcal{T} := \{1, \ldots, N^R\} \cup \{p\text{-tank}\}$ of indices is used. Most of the variables are nonnegative (continuous) flow variables

$$m_{pr} = M_p n_{pr}, \quad p \in \mathcal{P} := \mathcal{L} \cup \mathcal{G} := \{A, B, C, P\} \cup \{G_1, G_2, G_3\} \quad (3.66)$$

describing the total mass flow of product p or gas g into or out from node r, a reactor or the product tank. We distinguish between the liquids $\mathcal{L}$ and the gases $\mathcal{G}$ because they are subject to different topologies. While the variables m have the dimension tons/hour the variables n are in kmol/hour; they are coupled by the molecular masses M_p. Other variables are the temperature T_r and pressure p_r^P in reactor r, the stirring energy e_r used in reactor r, and as auxiliary variables, the weight fractions w_r^p.

The valuable product, P, is produced by a system of reactors r. The reactors are connected according to free or fixed pattern (single chain, parallel chains of different lengths, parallel chains with connections) as shown below

or parallel chains with connections

$$\begin{array}{l}
\rightarrow \rightarrow \rightarrow \rightarrow \rightarrow \rightarrow \\
\quad \searrow \quad \searrow \\
\rightarrow \rightarrow \rightarrow \rightarrow \rightarrow \\
\rightarrow \rightarrow \searrow \\
\rightarrow \rightarrow \rightarrow \rightarrow \rightarrow \rightarrow
\end{array} \quad .$$

During the synthesis within a reactor r side-products are produced like A, B or G_1, and raw material remains. Liquid components are fed to one or several subsequent reactors or via filter to the product tank. For the gaseous components incidence tables control the flow between reactors. The gaseous outlet of some reactors can lead to the incinerator or can be fed back to a reactor.

Mass balances for the reactors

The input side of reactors is described by

$$n_{pr}^i = I_{pr}^S n_{pr}^S + \sum_{s \in \mathcal{R} | I_{psr}^l = 1} x_{psr} n_{ps}^o, \quad \forall r \in \mathcal{R} \cup \mathcal{T}, \quad \forall p \in \mathcal{P} \tag{3.67}$$

For a fixed topology of reactors the (binary) incidence table I_{pr}^S describes whether reactor r is connected to a supply tank of product p; the variable n_{pr}^S describes the flow of product p from the supply tank to reactor r. If the topology is free we just set $I_{pr}^S = 1$ for all combinations.

The second term describes the flow of product p from all possible reactors $r_s \in \mathcal{R}$ to reactor r. The fractions, $0 \leq x_{psr} \leq 1$, distribute the output flow from a certain reactor to other subsequent reactor. Conservation of total flow is enforced by

$$\sum_{d \in \mathcal{R} | I^I_{prd} = 1} x_{prd} = 1, \quad \forall r \in \mathcal{R} \cup T, \quad \forall p \in \mathcal{P} \tag{3.68}$$

For a fixed topology with either single or several unconnected parallel chains we have $x_{prr+1} = 1$, and $x_{prd} = 0$ for all other combinations. For all products p within a stream from s to r we enforce the pooling condition

$$x_{psr} = x_{p'sr} \tag{3.69}$$

which expresses the conservation of composition. The synthesis of P needs a catalyst to be fed to the reactors. Since none of the catalyst flowing through the reactors is consumed we can approximately describe the flow of the catalyst by only one continuous variable, m^{CAT}. The mass of the catalyst is separated from the product outflow and reused again in the reactors.

How many kmol/h of substance p will leave the reactor r is described by

$$n^o_{pr} = n^i_{pr} + \sum_{p' \in \mathcal{P} | p' \neq p \wedge S_{pp'} \neq 0} S_{pp'} \Delta n_{p'r}, \quad \forall p \in \mathcal{P}, \quad \forall r \tag{3.70}$$

with

$$\Delta n_{p'r} := n^o_{p'r} - n^i_{p'r}, \quad \forall p \in \mathcal{P}, \quad \forall r \tag{3.71}$$

From the throughputs, $\Delta n_{p'r}$ of products p' in reactor r, with known reaction scheme represented by the stoichiometric coefficients $S_{pp'}$, we can derive the production and loss terms of product p. The amount of product p' leaving reactor r is the amount which has gone into the reactor and the amount which is produced in the reactor. The amount $\Delta n_{p'r}$ produced depends on the reaction rates $r_{p'r}$ and the volume of the reactor V_r:

$$\Delta n_{p'r} = n^o_{p'r} - n^i_{p'r} = r_{p'r} V_r, \quad \forall p' \in \mathcal{P}, \quad \forall r \tag{3.72}$$

In the current case, C is consumed by the production of P and its byproduct B – i.e. S_{CP} and S_{CB} and nonzero. Product A flows from a certain reactor to all possible subsequent reactors, but it is also produced as a by-product of the P-synthesis, namely during producing B and G_1 – i.e. S_{AG_1} and S_{AB} are different from zero.

G_1 is produced in a undesired reaction of G_2 and G_3. So the amount leaving the reactor consists of the amount flowing in and the amount produced depending on reaction rates r_{G_1r} and reactor volume V. The gaseous basic chemicals G_2 and G_3 are consumed within the reaction. The consumption depends on the amount of P, B and G_1 produced – i.e. we need to consider S_{G_3P}, S_{G_3B} and $S_{G_3G_1}$.

Reaction rates and weight fractions

The amount of P, B and G_1 produced in reactor r depends on the reaction rates. These, in turn, are functions of temperature, density, concentration of catalyst, and other parameters describing the chemical synthesis.

In the following the reaction rates are formulated for the substances which are synthesized or which decay. While for elementary reactions the rates are of the form

$$r_{AB} = k_0 [A][B] e^{-\frac{E}{kt}} \tag{3.73}$$

where $[A]$ and $[B]$ denote the concentrations of compounds A and B, the reaction rates in the current scheme are more complicated and depend especially on interpolated functions for density, viscosity, etc. The reaction rates r_{Pr} and r_{Br} for the P and B synthesis are computed by

$$r_{Pr} = r_{Pr}^{(1)} - r_{G_1r}^{(2)}, \quad r_r^B = C_1 m^{CAT} f_5(c_r^l) \tag{3.74}$$

where c_r^l is an auxiliary variable involved in the interpolation of the reaction kinetics, $r_{Pr}^{(1)}$ is computed in formula (3.79), and C_1 is a constant. The rate $r_{G_2r}^{(2)}$ describing the consumption of G_2 can be calculated as the sum of the rates

$$r_{G_2r} = r_{Pr} + r_{Br} + r_{G_1r} \tag{3.75}$$

and $f_n(x)$ using the catalyst data $P_1 = 25\,700$ and $P_2 = 400$ is defined as

$$f_n(x) := h_n(x) e^{-P_1 h_1(x)}, \quad h_n(x) := \frac{x}{(1 + P_2 x)^n} \tag{3.76}$$

Since c_r^l is of the order of $5 \cdot 10^{-5}$, by substituting $x = s(y)$ and exploiting the partial fraction relation

$$x = 5 \cdot 10^{-5}(1+y), \quad \frac{1+y}{\alpha+\beta y} = \frac{1}{\alpha} + \frac{\alpha-\beta}{\alpha} \frac{y}{\alpha+\beta y} \qquad (3.77)$$

(3.76) can be replaced by the numerically more stable expression

$$f_n(x) = g_n(y) := \frac{5 \cdot 10^{-5}(1+y)}{(1.02+0.02y)^n} e^{-\frac{1.825}{1.02}} e^{\frac{1.825}{1.02} \frac{y}{1.02+0.02y}} \qquad (3.78)$$

Finally we calculate the rate, $r_{G_1 r} = r_{G_1 r}^{(1)} + r_{G_1 r}^{(2)}$, of the synthesis of G_1

$$\begin{pmatrix} r_{G_1 r}^{(1)} \\ r_{G_1 r}^{(2)} \\ r_{Pr}^{(1)} \end{pmatrix} = m^{CAT} \begin{pmatrix} C_2 c_r^l f_2(c_r^l) \\ C_3 c_r^l c_r^P e^{-n_{Gr}^i} \\ C_4 f_3(c_r^l) \end{pmatrix}, \quad c_{Pr} = \frac{n_{Pr}^o}{V_r} \qquad (3.79)$$

The concentration, c_{Pr}, of P in reactor r is directly available, c_r^l can be calculated only implicitly using the nonlinear Arrhenius equation

$$g(c_r^l) := c_r^a h - 2(c_r^l) + 2000 m^{CAT} f_0(c_r^l) - c_{G_2 r} c_r^a (1+P_2 c_r^l)^2 = 0 \quad (3.80)$$

where c_r^a is the gas–fluid exchange coefficient in reactor r depending on the energy e_r,

$$c_r^a = C_5 r_r^{0.7} (1000\rho r)^{0.27} (0.001\eta r)^{-\frac{5}{6}} \qquad (3.81)$$

The concentration, $c_{G_2 r}$, of G_2 at the phase boundary

$$c_{G_2 r} = \frac{\alpha_r}{22.4} pG_2 r \qquad (3.82)$$

depends on the gas solubility, α_r. Both α_r and the viscosity η_r in reactor r are polynomials of third order in two variables

$$\begin{aligned} \alpha_r &= C_6 + (\mathbf{v}^T \mathbf{H}_1 \mathbf{v}) \\ \eta_r &= C_7 + (\mathbf{v}^T \mathbf{H}_2 \mathbf{v}) \end{aligned}, \quad \mathbf{v}^T = (w_{Pr}, w_{Cr}) \qquad (3.83)$$

where $\mathbf{H}_1$ and $\mathbf{H}_2$, as well as $\mathbf{H}_3$ and $\mathbf{H}_4$ used below, are constant matrices of appropriate dimensions. The mean density ρ_r in reactor r is a polynomial of fourth order – i.e.

$$\rho_r = C_8 + (\mathbf{u}^T \mathbf{H}_3 \mathbf{u})(\mathbf{u}^T \mathbf{H}_4 \mathbf{u}), \quad \mathbf{u}^T = (w_{Pr}, w_{Br}, w_{Cr}, T_r) \tag{3.84}$$

Finally, the weight fractions depend nonlinearly on the molecular masses according to

$$w_{pr} = m_{pr}^o \bigg/ \sum_{p' \in P_p} m_{p'r}^o, \qquad \begin{matrix} \forall r \in \mathcal{R} \\ \forall p \in \mathcal{P} \\ \mathcal{P}_p := \{p' \in \mathcal{P} | p' \neq p\} \end{matrix} \tag{3.85}$$

Discrete features of the model

The dominant discrete feature is the requirement that the flow rates between reactors do not become arbitrarily small. This is guaranteed by binary variables δ_{sr} indicating that reactor s has a connection to reactor r and the constraints

$$C_{sr}^{\min} \delta_{sr} \leq x_{psr} \leq C_{sr}^{\max} \delta_{sr}, \quad \forall p \in \mathcal{P}_{sr} \tag{3.86}$$

Three additional inequalities ensure that each reactor has at least one inflowing and one out-flowing stream, and that the number of subsequent reactors fed by reactor r does not exceed a maximum number N^{SR} – i.e.

$$\sum_{r' \in \mathcal{R}} \delta_{r'r} \geq 1, \quad 1 \leq \sum_{r' \in \mathcal{R}} \delta_{rr'} \leq N^{SR}, \quad \forall r \in \mathcal{R} \tag{3.87}$$

A similar constraint is used to enforce that at least one reactor is connected to the filter. The logical constraint

$$\delta_{Arr'} = \delta_{Crr'} \tag{3.88}$$

is used to guarantee that the liquids A and C are not solely used.

Finally, we need to select the size of the reactors. The mathematics is almost identical to the selection of the size of pipelines in subsection 4.1, and is not repeated here.

The objective function

We consider four alternative objective functions: The first one is to maximize the selectivity of P defined as reaction rate r_p of P over reaction rate, r_{G_2}, of G_2, the most expensive raw material. First experiments with this objective function showed that solutions are produced with high selectivity but only for a low amount of P produced. Thus in this objective function scenario we require that a certain amount of P has to be produced. The second objective function maximizes the total mass of P. The third objective function minimizes the variable cost to produce P while guaranteeing that a certain minimal amount of P is produced. A fourth objective function minimizes the total consumption of stirring energy.

5.2. Solution approach

The convergence of the problem depends critically on appropriate initial values required to solve the NLP problems. So, in the beginning we often experienced divergence. The numerics improved when variables and constraints were rescaled. It also became necessary to apply bounds on some of the process variables (e.g. temperature, pressure, etc.) to keep the values in physical realistic ranges. A special example of scaling is related to the quantity c_r^l used as an argument of (3.76) in many places. Finally, GAMS supported the computation of useful initial values by minimizing the violation of certain equations. Among the most difficult one is (3.80). In this case the minimization of violation variables $\mathbf{v}^+$ and $\mathbf{v}^-$ in the relaxation $\mathbf{g}(c) = \mathbf{v}^+ - \mathbf{v}^-$ of $\mathbf{g}(c) = 0$ provides good initial value in short time. For fixed topology, with appropriate initial values, the NLP problems is solved in a few minutes. This approach is more typically for a production planning system. Current initial values are stored and reused in the next production planning runs in daily life. The real design problem and the full MINLP approach is used by experts only to investigate new situations.

5.3. Case study results

The following description of optimization results are based on the analyses of a typical reference scenario provided by the client. If not mentioned otherwise the feedback of products from the sixth reactor to the first is set to zero. For fixed topology (a chain of six reactors) we get within less than one minute of CPU time:

- The total amount of stirring energy has been kept constant but the optimizer was free in distributing it to the six reactors. The input mass flow of the chemicals were not kept constant compared to the reference solution. The optimization yielded a slightly larger consumption of G_3, G_2 and C and an increasing (from the very first reactor) need of stirring energy. The mass fraction of P and the mass of P produced were increased by about 1%.

- Here we wanted to minimize the total stirring energy while a minimum production of P was required in order to compare the results with the reference solution. The results show a slightly higher consumption of G_2, for G_3, comersely, while the total energy need was about 8% lower than for the standard scenario.

- Distribution of the stirring energy with a constant amount of total energy but in contradiction to scenario $E1$ the input of chemicals was set to be the same as in the reference scenario. The results show a distribution of the stirring energy increasing from the first reactor whereas all other output values did not improve.

- Here we wanted to investigate the influence of the feedback of the products from the last to the first reactor. The results show that the amount of P produced increases if one requires a certain feedback (e.g. 10% or 20%). Unfortunately, this gain gets lost because of the feedback so that the effective amount of P is reduced compared with the reference solution. The optimal solution is found for the case of no feedback.

The optimization model has been embedded into an attractive and easy-to-use userinterface. It helps the client in his daily production planning duties to adjust his plant immediately to current needs – i.e. changes in costs, capacities' fluctuations, or attributes of orders. The tool supports the design phase and helps to lay out cascades and connections of a system of reactors. Here, the client sees the benefit being able to compare variants proven optimal or at least of known quality. The new designs save raw material, minimize waste material and increase the capacity of the reactor system. In the lay-out phase the tools support design and other changed constraints.

VI. CONCLUSIONS

In this chapter, mixed-integer nonlinear optimization has been considered as an approach to solve complex production planning and design problems. The problems discussed are very demanding in terms of the mathematical modeling and appropriate tuning of the algorithms. In all cases, special heuristics were constructed to provide reasonable initial values to the solver.

In the first case it was possible to substitute the nonlinear terms by equivalent linear terms involving binary variables. In the second and third problem, dominated by pooling problems, initial guesses for the fractional composition of the multi-component streams could be derived from a simplified linear model. A homotopy method is used in the third problem by solving a sequence of sub-models of increasing complexity (LP, NLP-1, NLP-2, MINLP) exploiting the results of the previous ones. The third problem is solved in sequence of sub-models formulated in GAMS, exploiting the results of the previous one – i.e. we use a homotopy method. A simple linear model provides initial values for two simple nonlinear submodels. In the fourth model scaling was very important. Solving an auxiliary problem in which an artificial objective function measuring the violation of certain nonlinear constraints was minimized provided excellent initial guesses.

Future direction regarding the first problem will focus on special branching rules and cuts to improve the gap. For the current application this is not a problem because only a few products required constraints across periods. The third and fourth problem helped to accumulate experience in solving MINLP problems. However, the lesson to be learned is that each MINLP problem is different and requires special treatment and techniques. One common feature seems to be the problem of getting good initial values to start the solver which, according to our experience, can be overcome by homotopy techniques.

The heterogeneous approaches to solve the problems indicate that mixed-integer nonlinear optimization is an area under continual development. It has proven itself as a useful technique to reduce costs and to support other objectives, and it certainly has much to offer for the future. MINLP is another example showing that mathematical methods and techniques can support human inventiveness and decisions. In particular, they can ensure that less intuitive solutions are not lost, and can provide a quantitative basis for decisions and allow us to cope most successfully with complex problems.

Notes

1. Thanks are directed to Christian Timpe (section III) and Norbert Vormbrock (section IV and V), who provided material or joined the author in the project work. This chapter benefitted greatly from suggestions and comments provided by Anna Schreieck and Beate Brockmüller.
2. Although the problem instance specified by the data leads to a one-to-one relation between modes and products, the coupling $p_{rpt} \leq \Sigma_m R_{mp} m_{rmt}^D$ with

production rates R_{mp} holds for any mode–product relation. Note that in M1 $r \Leftrightarrow i$, and $t \Leftrightarrow k$.
3. The model is formulated and solved by PIMS (Module PPIMSXX and XPIMS) by Bechtel Corp. (Houston, USA, now ASPEN Tech).

References

Ashford, R.W. and R.C. Daniel (1987) LP-MODEL XPRESS-LP's model builder', *Institute of Mathematics and its Application Journal of Mathematics in Management*, 1, 163–76.

Ashford, R.W. and R.C. Daniel (1991) 'Practical Aspects of Mathematical Programming', in A.G. Munford and T.C. Bailey (eds), *Operational Research Tutorial Papers* (Birmingham: Operational Research Society), 105–22.

Bazaraa, M.S., H.D. Sherali and C.M. Shetty (1993) *Nonlinear Programming: Theory and Algorithms* (New York: Wiley).

Brockmüller, B. and L. Wolsey (1995) 'BASF Problem 1 Modelling – Production Scheduling with Batch-Sizing and Storage Restrictions', report of ESPRIT project PAMIPS, Task 1.4 8/1995, BASF-AG, ZX/ZC-C13.

Brooke, A., D. Kendrick and A. Meeraus (1992) *GAMS – A User's Guide (Release 2.25)* (Danvers, MA: Boyd & Fraser).

Duran, M.A. and I.E. Grossmann (1986) 'An Outer-Approximation Algorithm for a Class of Mixed-Integer Nonlinear Programs', *Mathematical Programming*, 36, 307–39.

Floudas, C.A. (1995) *Nonlinear and Mixed Integer Optimization* (Oxford: Oxford University Press).

Kallrath, J. and J.M. Wilson (1997) *Business Optimisation Using Mathematical Programming* (London: Macmillan).

Kallrath, J., Y. Pochet and S. Raucq (1994) 'BASF Problem 1 Modelling – Production Network Planning System', report of ESPRIT project PAMIPS, Task 1.1 10/1994, BASF-AG, ZX/ZC-C13.

Leyffer, S. (1993) Deterministic Methods for Mixed Integer Nonlinear Programming', PhD thesis, Department of Mathematics and Computer Science, University of Dundee.

Nemhauser, G.L. and L.A. Wolsey (1988) *Integer and Combinatorial Optimization* (New York: John Wiley).

Viswanathan, J. and I.E. Grossmann (1990) 'A Combined Penalty Function and Outer-Approximation Method for MINLP Optimization', *Computers and Chemical Engineering*, 14, 769–82.

4 Lube Production Scheduling Model at the AGIP-Petroli's Livorno Refinery

Sergio Barbariol, Mauro Lusetti,
Marco Mantilli and Mauro Scarioni

I. INTRODUCTION

The production scheduling process is a fundamental phase in order to reduce production costs and final product demand optimization. The petrochemical industry, characterized by huge amount of material and by an easy continuous process in terms of mathematical modeling, is very interesting for automatic scheduling tools.

The advantage of efficient automatic tools can be summarized as:

- high speed in decision phase
- scenario and 'what-if' analysis
- creation of a quality index, that can judge the quality of the scheduling in analytical mode
- simplifying the scheduling problem.

All the items support excellent characteristics for:

- schedulers who are interested in application interchangeable techniques
- casual users looking for optimization tools that are cheap to learn and easy to apply.

The methodology implemented is based on mixed-integer programming (Ballintijn, 1993). The linear programming (LP) methodology has been used for a long time for planning applications, where nonlinear problems are solved using recursive methods in form of successive linear approximation (Kallrath and Wilson, 1997; see also Chapter 7 in this volume). Recently the application to operative scheduling has required integer variables.

The main objective of this project is to automate and optimize the scheduling process of lube, the main production of the AGIP-Petroli's Livorno Refinery.

II. PROBLEM DESCRIPTION

The refinery runs an integrated hydroskimming-type cycle for the production of fuels. The fuel yields high-quality products: LPG, Kerosene, fuel oils, and dearomatics feedstock for the production of aliphatic solvents.

This chapter describes the lube production plant with a Solvex-type cycle that takes advantage of the more advanced technology.

- The plant outputs are transformed into a range of lube bases, waxes, and bitumen for the national and international market.
- The complex and integrated productive cycle is able satisfy the market demand of about 500 000 ton/year of lube bases.

The lube plants consist of

- *Vacuum distillation unit* (VPS) that processes residue from atmospheric distillation in order to produce vacuum gasoils, which are sent to the desulphurization unit in the fuels section; side stream, which make up the feedstock for the refining unit; and vacuum residue which is used in part as feed for the deasphalting unit, and in part as a component in fuel oil blending.
- *Deasphalting unit with propane solvent* (PDA) which removes a heavy base from vacuum residue which is used as feedstock to the refining unit with furfural solvent.
- *Refining units with furfural solvent* (FT1/2) which remove the aromatic components from the bases used for finished lubricant production and increase their viscosity stability in conditions of temperature change.
- *Dewaxing units with solvent methyl, ethyl, ketone and toluene* (MEK1/2) which remove waxes from lubricant bases to improve their pour point characteristics.

Other plants related to lube process are present in the refinery, but not related to the project. The output of a plant is directly fed into other plants or can be stored in tanks. This stock operation is particularly important for the vacuum unit (VPS) that produce contemporaneously four vacuum residues. The tank should be used to keep the remaining two residues that are not used from the refining unit (FT1/2).

- We define two plans in *cascade condition* when the output stream of a refining unit is directly linked to the input stream of a dewaxing unit.
- The cascade condition improves the final product quality.
- At each plant we define, for each time slice, the configuration of the unit, defining at the same time the input stream, the blend needed for this stream and the output stream.

Each configuration is characterized by the following:

- feasibility of the configuration for the plant
- minimal feed charge
- maximal feed charge
- plant rate
- minimal length of the configuration
- set-up cost for the configuration.

The plant scheme and the connections are schematically represented in Figure 4.1.

The streams have been posed in three logical groups, following the classical use of the refinery; for each stream the following attribute should be defined, if available and feasible:

Figure 4.1   Logical scheme in the plant

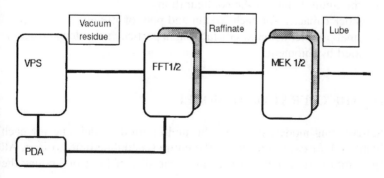

- stock capacity and mode
- stock cost
- downgrade cost
- no demand satisfaction cost.

The target of the scheduling system is composed of five different factors:

(a) **Set-up costs** – related to set-up of each plant, depending on the closing and opening configuration.

(b) **Cascade stream costs** – the cost of the cascade stream is defined by the contemporary configuration of a refining unit with a dewaxing unit of the same stream as the output of FT and input for MEK. The shutdown of a unit is considered as a stream, so the contemporary shutdown of a refining unit and a dewaxing unit is considered a cascade.

(c) **No-show costs** – this is the quantity of the demand of final product that is nonsatisfied, multiplied by the delay in terms of time-slice.

(d) **Stock costs** – for each stream a stock cost is defined for each ton of weight of product.

(e) **Downgrade cost** – instead of being stocked, the vacuum residue could be downgraded as fuel. With such operations the economical value of the product decreases, then we need to take into account this value loss for each ton of products.

The scheduling optimization should take into account five technical constraints:

- minimum and maximum charge for unit
- feasibility of the configuration, for each unit
- minimum length of the configuration
- stock capacity, for each stream and pool of stream
- final product demand satisfaction, in particular the demand generated by shipment.

III. THE OPERATIONAL MODEL

Set-covering models compose the multi-period model, one for each plant and for each time-slot of the entire scheduling time horizon. All the entities are scaled, regarded as a time-slice of 1 day or 8 hours; the scheduling should be done by assigning a slot (or a set of slots) to a asset, identified by the name of the product in output.

3.1. Input data

Final product demand, split by shipment system (ship or truck), and scheduled in the time horizon, is divided in two parts – the first that should be satisfied before the due date, and the second that could be delayed. The later part is balanced by a variable that indicates, day by day, the quantity of demand not satisfied.

Plant yields (furfural refining, dewaxing, and deasphalting) are defined by the ratio between the quantity of output and the quantity of feed for each configuration. A link matrix is defined between the units' input and the final product. The matrix elements consider also the max rate for the blending of different components.

- minimum and maximum rate for each unit and each configuration
- initial configuration of the unit
- initial stock for each intermediate and finished product
- minimum value of the finished product stock at the end of the scheduling
- forced daily configuration of the unit, useful for a better control of shutdown and special-product production
- minimum and maximum stock value of intermediate and finished products
- minimum batch length related to configuration of furfural refining and dewaxing unit and to the unique value of deasphalting and vacuum unit
- vacuum unit production, in terms of quality and quantity for each configuration feasible for input crude oil
- furfural refining and dewaxing configuration links in order to define the cascade.

3.2. Variables

Real:

- **stock quantity** of each product, vacuum residual, refining, and lube
- **downgraded intermediate product quantity** for each time-slot, related to stock problem for vacuum residual
- **process unit input quantity**, for furfural refining unit and for dewaxing unit

- **unsatisfied demand**, defined as the difference between total demand and production before the time-slot
- **configurations not in cascade condition**, defined as furfural refining unit configuration not related to the dewaxing configuration.

Binary:

- **the configuration of each unit for each time-slot**; special ordered set variables type 3 (OSL, 1992) on the unit and time-slot related to vacuum, deasphalting, furfural refining, and dewaxing units
- **the set-up** for unit and time-slot; since we consider a limited number of configurations, vacuum and deasphalting unit binary variables refer to input configuration only, furfurals refining and dewaxing unit variables refer to input and output configuration, because they represent a significant difference.

Constraints:

- **material balance** for intermediate and finite products
- **the balance of the vacuum residue**, which considers the following quantities:
  - the vacuum and deasphalting unit production
  - the input quantity for the furfural refining and/or deasphalting unit
  - the stock
  - the downgraded quantity
  - input charge for dewaxing unit for particular configurations.
- **The balance of refining** refers to the following quantities:
  - the output of furfural refining unit
  - the input of dewaxing unit
  - the stock.
- **The balance of lube** includes the quantities:
  - the dewaxing unit production
  - the stock
  - the demand
  - the unsatisfied demand.
- **minimum and maximum input value** of each unit and configuration
- **limit to unique configuration** for unit and time-slot
- **minimum unit configuration length** – vacuum, deasphalting, furfural refining, and dewaxing unit configuration has a minimum

length defined in terms of time-slot; this constraint forces us to maintain a configuration for a minimum number of time slots, in order to avoid assets being maintained for very short terms.

- **Consistency of *number-of-cascade* variable** – For each group of cascade configurations we define the number-of-cascade variable that indicates, for each time-slot, the difference between the number of furfural refining units and the number of dewaxing units.

3.3. Bounds and fixed variables

- **Maximum limit for the stock** of each product: vacuum residue, refining and lube. These products are stocked in shared tank and the bound refers to the strategic limit of the stock, rather than the overall limit regarding the operational limit.

The refining has an added overall limit for the total amount.

- **Forced configuration.** In some cases, especially for shutdown or special-product production, the operator needs to fix a certain unit configuration.

In this case, a fixed bound to 1 is defined for that time-slot and that unit. The initial unit configuration should be fixed as well.

Depending on external data, some configuration for a particular-time slice could also need to be fixed.

The constraints listed here were considered accurate enough to describe the petrochemical process, and the results of the optimization accurate enough for process scheduling. But the model is too complex and time-consuming to the used in practice.

In fact, the project aims at providing a satisfactory solution in less that 10 minutes of CPU time. Then we introduce an advanced modeling technique able to reduce the solution time even further.

IV. ADVANCED MODELING

New constraints (Johnson, 1980) can be found in order to reduce the feasibility region and the gap between the relaxed model and the integer solution.

This reduction can be performed at the modeling level by managing some logical condition (Nemhauser and Wolsey, 1988; Ciriani and Gliozzi, 1994).

The set-up logical variables offer the better field in which to operate, because they contribute heavily to integer infeasibility of the optimal relaxed model solution. A first step is the definition of the (logical) variable G_j that indicates the end of the configuration of the j unit, so when a S_j occurs then C_j should take place one time-slice later

$$\sum_j S_{j,t-1} + X_{j,t-1} = \sum_j C_{j,t} + X_{j,t} \quad \forall t$$

where:

$X_{j,t}$ assumes 1 value if configuration of j unit is active at time t, 0 otherwise

$S_{j,t}$ assumes 1 value if configuration of j unit sets up at time t, 0 otherwise

$C_{j,t}$ assumes 1 value if configuration of j unit ends at time t, 0 otherwise

Since a unit cannot have more $C_{j,t}$ at a defined time-slice than $S_{j,t}$ in the previous time-slice, then we introduce the constraint:

$$\sum_{k<t} S_{j,k} \geq \sum_{k\leq t} C_{j,k} \quad \forall t, j$$

With the same logical rule we can define that $X_{j,t-1}$ is always greater than $S_{j,t}$

$$X_{j,t-1} \geq S_{j,t} \quad \forall t, j$$

In addition, we should compute how many time-slots of each finite product we need to schedule before each milestone. So we add a balance constraint that sets the production in terms of tons.

The new constraints:

- fix a set-up before a time-slice for an active and noninitial configuration, for furfural refining and dewaxing unit
- fix a number of time-slot allocations enough to cover the ship demand, calculated with the following rule:
  - for lube it will assume the higher output of a dewaxing for time-slice

– for refining it multiplies the quantity of lube needed by the higher yield of the dewaxing unit and takes the higher output of a furfural refining unit.

V. OPTIMIZATION STRATEGY

The model is formulated by the language IBM AIX EasyModeler/6000 (EasyModeler, 1994) and solved by IBM AIX Optimization Subroutine Library/6000 (OSL, 1992). Primal simplex finds the optimal solution of the relaxed model in less than 3 minutes of CPU time.

Tests performed with the mixed-integer programming preprocessing (Ciriani, 1993) of OSL were not able to generate strong cuts in a reasonable CPU time. The general strategy of the branch-and-bound is giving priority to a greedy solution. To reduce the solution time, the priority of the branch has been changed. The implemented strategy gives priority to the last unit in the process, so high priority for dewaxing unit, less for the furfural refining unit, less for deasphalting and for the vacuum unit.

The time dimension gives high priority to the first time-slice, less to second, and so on. The scheme used starts from an easy concept: the critical scheduling is the beginning of the time horizon, because we have constraints in stock. At the end of the time horizon we probably have more degrees of freedom, because we can schedule different configurations and we can manage use for feeding and production of the various intermediate and finished products.

VI. COMPUTATIONAL RESULTS

One of the major targets of the project was to solve a real-life problem, based on 15 time-slots in less than 10 minutes of IBM RISC System/6000 CPU time.

Usually a sub-optimal solution is accepted when the gap of the objective function value is *satisfactorily* narrow and the solution time remains within the predefined boundaries.

A solution is considered *satisfactory* when the number of unit set-ups is a minimum.

Test instances, based on time series, show the feasibility of the target, sometimes with a solution time of less than 5 minutes. This happens also in a critical situation, with shutdown of unit very close to the start of scheduling horizon usually considered a very hard problem.

Table 4.1   Real data tests.

Test	Row	Cols	Integer vars	Lp solution		1st integer solution		Set-up cost
				Obj.	secs	Obj.	secs	
Istance 1	4064	5936	504	8192	155	20804	416	10000
Istance 2	3974	5672	480	29175	82	36897	212	10000
Istance A	3974	5672	480	18066	124	20728	473	10000

Table 4.1 shows three real data tests.

VII. MODEL DEVELOPMENT AND USER INTERFACE

The mathematical model has been formalized using IBM AIX EasyModeler/6000 language, strongly using the selecting row and columns' generation capability, reducing the matrix size and inserting cut before the branch-and-bound algorithm (Scarioni and Sciomachen, 1995).

The model set are generated in a fixed format and scaled in terms of time-slice duration. The data are calculated in real-time, during the instance generation, because the application allows the user to change the time-scale. MIMI is the standard scheduling tool used in the refinery for all other scheduling applications.

We then implemented an interface to MIMI for model formulation and solution. Figure 4.2 shows an example of the interface used to input ship finite product demand, giving the arrival gate, ship name, product, quantity and availability, and needed date of the product in the tank.

Through this interface the user can define a time horizon as wide as the scheduling horizon. The model's data are prepared and scaled with required measure unit. Then the data are transferred to EasyModeler and the model is solved. At last the interface loads the solution into MIMI.

Figure 4.3 shows a Gantt chart generated by MIMI's Planning Board.

Figure 4.3 reports the configuration of a single unit for a time horizon of 20 time-slots. Starting from the top vacuum feed the planning board follows with vacuum residue production, deasphalting confi-

Figure 4.2   Data input screen for the ship demand of lube

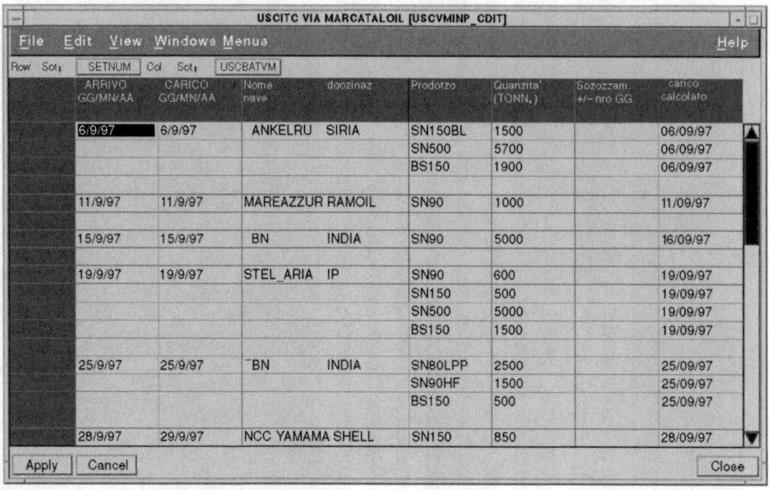

ARRIVO GG/MN/AA	CARICO GG/MN/AA	Nome nave	doozinaz	Prodotzo	Quanzita' (TONN.)	Sezozzam +/- nro GG	carico calcolato
6/9/97	6/9/97	ANKELRU	SIRIA	SN150BL	1500		06/09/97
				SN500	5700		06/09/97
				BS150	1900		06/09/97
11/9/97	11/9/97	MAREAZZUR	RAMOIL	SN90	1000		11/09/97
15/9/97	15/9/97	BN	INDIA	SN90	5000		16/09/97
19/9/97	19/9/97	STEL_ARIA	IP	SN90	600		19/09/97
				SN150	500		19/09/97
				SN500	5000		19/09/97
				BS150	1500		19/09/97
25/9/97	25/9/97	BN	INDIA	SN80LPP	2500		25/09/97
				SN90HF	1500		25/09/97
				BS150	500		25/09/97
28/9/97	29/9/97	NCC YAMAMA	SHELL	SN150	850		28/09/97

note: The English correspondence to the Italian terms used in the header is:

ARRIVO GG/MM/AA	ARRIVAL DD/MM/YY
CARICO GG/MM/AA	SHIPLOAD DD/MM/YY
destinaz.	destination
Prodotto	Product
Quantità	Quantity
Spiazzam. nro GG	Delay number of days
carico calcolato	Computed shipload

guration, furfural-refining configuration, dewaxing configuration, and downgrade product quantity (equal to zero in this example). In the graph below the Gantt chart we can see the profile of the stock, with a dotted line when the balance is negative due to the rounding difference between the LP computed and the real system. A cascade occurs when furfural refining and dewaxing configuration have the same color.

VIII. CONCLUSIONS

Valid benefits come from better planning, from ability in scheduling and decision-making, from improved productivity, and from more timely and accurate information.

88

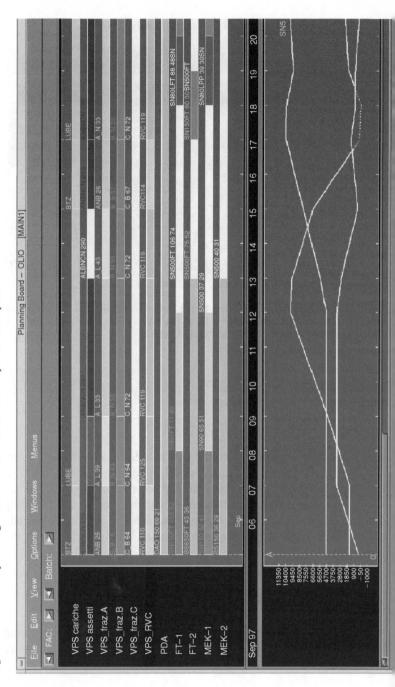

Figure 4.3 MIMI's planning board and the result of the optimization phase

Major credits can be realized from these areas:

- fewer inventories required and reduced working capital
- increased productivity of planning and scheduling staff
- longer run lengths, with effective higher capacity and improved yields
- less quality giveaway and closer adherence to product specification with the maximization of the cascade between the units
- improved monthly planning and lube optimization
- better ability to handle operation upsets and unplanned unit shutdowns
- improved customer satisfaction from just-in-time and more reliable scheduling
- improved coordination and communication among planning, scheduling process unit operations, and oil movements
- faster responsiveness to market changes.

The scheduling model can also be used to support the following enterprise functions:

- inventory management and tracking (tankages analysis)
- more 'real-time' decision making ability.

The lube model was tested and activated for real production, while further extension of the model and scope to the overall lubes and waxes scheduling proceeds.

References

Ballintijn, K. (1993) 'Optimizing in Refinery Scheduling: Modeling and Solution', in T.A. Ciriani and R.C. Leachman (eds), *Optimization in Industry* (New York: Wiley).

Ciriani, T.A. (1993) 'Model Structures and Optimization Strategies', in T.A. Ciriani and R.C. Leachman (eds), *Optimization in Industry* (New York: Wiley).

Ciriani, T.A. and S. Gliozzi (1994) 'Algebraic Formulation of Mathematical Programming Models', in T.A. Ciriani and R.C. Leachman (eds), *Optimization in Industry*, vol. 2 (New York: Wiley).

EasyModeler (1994) IBM AIX EasyModeler/6000, User Guide, Release 2.0, SB13-5249, IBM Corporation, Rome.

Johnson, E.L. (1980) 'Integer Programming, Facets, Subadditivity, and Duality for Group and Semi-Group Problems', *SIAM*.

Kallrath, J. and J.M. Wilson (1997) *Business Optimization Using Mathematical Programming* (London: Macmillan).

Nemhauser, G.L. and C.A. Wolsey (1988) *Integer and Combinatorial Optimization* (New York: Wiley).

OSL (1992) Optimization Subroutine Library Release 2, Guide and Reference, SC23-0519, IBM Corporation, Armonk.

Scarioni, M. and A. Sciomachen (1995). 'Fast Prototyping for Optimization Models in Production Planning', in A. Sciomachen (ed.), *Optimization in Industry*, vol. 3 (New York: Wiley).

5 Practical Modeling

Larry Haverly

I. BEHAVIOR OF MODELS

Successful practical application of a mathematical model requires a good understanding of the behavior of the mathematical model by the person developing the model and an understanding of its limitations by the user. The model needs to be set up to reflect those essential features of the real situation being modeled. The distinction between the real situation and the model must be kept in mind, along with a recognition that the model is not the same as the real situation, but only our best representation of some important aspects of that real situation.

Since this chapter is devoted to linear and mixed-integer applications, I will direct my comments on behavior of models to these types. References to 'model' mean LP models, unless otherwise made clear. Many of these comments are made based on years of experience of building a wide variety of models for practical use, and the examples mentioned are from actual practice. I wish to thank the countless people, such as Beale (1988), whose ideas I have used or adapted to build practical applications.

Aspects of real applications may not always be linear. However, the Linear Programming (LP) technique is so powerful that there is an incentive to use LP where possible. It is amazing how often it has been possible to make linear approximations of somewhat nonlinear reality in a way which gives acceptable model behavior for practical purposes. It is also convenient for some nonlinear relationships to use extensions to LP such as Recursion (also known as Successive LP) or mixed-integer programming (MIP). An important example of the use of Recursion for the nonlinear pooling problem is given in Fieldhouse (1993) and Haverly (1979–1980).

An early paper on model behavior is Haverly (1977).

II. REPRESENTING DEMANDS AND PURCHASES

One area of nonlinear behavior involves interaction between selling prices and demands, and between purchase costs and amounts purchased. The price or cost may change based on the amounts involved. The amounts are typically row constraints in an LP model while the prices or costs are objective function values. There are algorithms, which will handle the objective/amount relationships nonlinear equations. However we frequently find that LP or MIP models give correct, or at least acceptable, behavior.

In this area we have often used step functions. An example would be to set the selling price for the first amount (A_1) at one price (P_1), the price for the next amount (A_2) at price (P_2), and so on. These require a few extra rows and vectors in the model but this is not usually a problem. However when we examine behavior, we not that an LP will not choose the A_1/P_1 option first if P_2 is a greater selling price than P_1. LPs always seek a higher revenue (or lower cost) over a lesser price (or higher cost). So behavior tells us that the step function will serve in an LP model only if P_2 is less than P_1, P_3 less than P_2 and if the reality is A_1/P_1 should be chosen ahead of A_2/P_2 and it ahead of A_3/P_3.

The reverse is true for purchases. The LP will choose the cheaper cost first, then the next higher cost, and so on. It needs C_1 less than C_2, and so on to have proper behavior if the real situation is that A_1/C_1 is a better choice than A_2/C_2.

An LP model will also tend to favor the corner solution, meaning that it will tend to sell maximum A_1 and may not have enough incentive to go to A_2. So there is a tendency for solutions to cluster at the break points, and users should be made aware of this behavior so they can fine-tune the mathematical results before using them in practice.

In some real situations, the amount/price relationships are not stepping functions but are continuous. That is, there is a smooth transition from one price–demand to the next. But in many practical situations the price–demand relationships are not definitively known so some approximation is acceptable as long as the driving force is correct. With proper choice of step sizes and judgment by the user, this apparent drawback can be minimized.

In general, it is good business practice to seek the highest-revenue customers first and then the next highest, and so on and stop increasing production when the marginal sale is less than the marginal cost of production. This is also the behavior of an LP and helps make LP so useful. But this is not always the full story. A high-price customer might require extra service or require extra costs so the true economics are less than they might appear from selling price alone. It may be possible to incorporate these costs in the model as reductions in net revenue. But some considerations are hard to model explicitly. For example, the highest-price potential customer might be just looking for a one-time purchase whereas a customer at a lower price might be a long-term steady buyer. The long-term buyer may be a lot more valuable than the one-time buyer. Blind faith in LP results would miss these kinds of important considerations. A possibility is to fix demand for certain customers to take them out of the LP choice so that their selling price plays no role in the LP solution.

Moreover many real situations involve real step functions. For example, each A_i/P_i combination may represent a single large customer or a group of smaller buyers in a particular channel of trade or geographic location. Regarding geographic location, if one sells on a delivered-price basis, then the selling price in the model may be reduced by the transportation cost to get a net-back price so that nearby customers have a higher net-back than more distant customers with the same delivered price. In these situations, the step functions are reality and not an approximation.

Price or cost steps can be the reverse of what was described above. For example, one may get a higher net price as more quantity supplied. This could occur in a buyers' market where new customers will pay more but you want to keep existing customers even at a slightly lower price. In this case, an LP model will not behave properly. We use MIP techniques to force A_1 to be met before A_2, and so forth. The technique of Special Ordered Sets (Beale, 1988) in most modern LP/MIP solvers such as CPLEX, XPRESS, OSL, and others may be especially useful and efficient for these types of step functions.

MIP will also be needed in situations where one is offered a lower price for purchases if they meet certain minimum volumes but price is higher below the minimum. LP does not behave properly in this situation.

III. REPRESENTING INVENTORIES

Another interesting area of practical concern, as related to model behavior, is representation of inventories. Where you have a multi-time period model and the model allows building of inventories in one time period but then uses them in a future time period, model behavior may closely represent reality if there are real differences in such factors as production costs, selling prices, demands, and capacity among time periods, and when the storage costs and capacities are properly reflected in the LP model. If sales have only maximum constraints and prices are lower in an early time period, the model may avoid sales in that period and put production into inventory for later sales. This may not reflect reality if in order to keep your customers you have to meet their demands in all time periods. However, by inclusion of minimum sales constraints in the low-price periods or by use of vectors which connect sales between periods, one can produce proper model behavior in the multi-period LP model.

A much more difficult behavior problem occurs when inventories are included in a single-period model. The same comments also apply to the last period of a multi-period model, but here the effect on the first period of the model may be minor. Normally, in a multi-period model one is planning to implement the decisions only for the first period and then review the situation later when it is time to make the second-period decisions. Even though only current decisions will be implemented, the multi-period model is used to incorporate a longer-range view when current decisions are closely connected to future choices.

The difficulties arise because the behavior of LP is to go to extremes. If something has a favorable objective, then LP wants to maximize it (or if it has an unfavorable objective, to minimize it). When we have final inventories in our model, if the value we put on one is greater than the production cost by even a tiny amount, then the LP will try to maximize that inventory. If the inventory value is even a tiny amount below production cost, then the LP will minimize that inventory. Often a user would want neither of these solutions but something in between. Also if the inventory value is set greater than the selling price of a maximum constrained demand, then the LP will put a product into inventory rather than sell it. This is often quite different than what a user would want to do since rejecting sales to build inventory is not normally considered good business practice. In the situation where the value on inventory is exactly the same as the production cost or the same as the selling price one gets unstable results. On one run the LP may give a solution which maximizes ending inventory and on the next minimizes it. This would be a flawed application and quickly lose credibility with users.

Good application practice is that the model must be structured so that it will behave in a way that gives the type of answers useful for the real situation. For the last-mentioned situation this can sometimes be achieved by fixing ending inventory levels or, occasionally by using a value for ending inventory which will give a desired behavior (such as a low value to minimize ending inventory level if that is the user's desire).

IV. REPRESENTING CERTAIN OPERATIONS

Certain aspects of a real situation may require mixed-integer programming (MIP) to achieve proper behavior. In some refineries (and other process plants) there may be physical restrictions that mean that one must choose between alternatives and not allow parts of choices (in other words, fractional solutions). There may be three tanks piped such that only two of them can be used at a single time. A simple integer constraint will model this condition. Automatic blending equipment may not work if a component to the blend is in small amounts (say under 5 percent). Again, a simple integer constraint can allow a component in a blend only if in amounts greater than the minimum.

Arrivals of crude oil at a refinery or shipment of products is normally in batch amounts. The crude may arrive in a large tanker holding hundreds of thousands or millions of barrels of crude oil. For planning selection of the best crude oils to buy and the average operation of the refinery over several months ahead, it is common to ignore batch sizes and deal with totals for the period and average daily amounts. But as the time span is shortened to a few days or weeks, model behavior when you deal with averages is no longer a good representation of the real situation, which must deal with distinct batches. Here MIP can play a major role to improve model behavior. Using some integer variables in our model we can require that a crude is used only in the amount that would be received by a shipload or a batch pipeline receipt. While a planning model might select four crude oils to run in a certain proportion on the average, the reality of storage facilities may be such that all four cannot be kept on hand all the time. Therefore spaced batch receipts might mean that only two are available during part of the period and at no time are all four available. A multi-period MIP model can include integer constraints that require each crude to be received and used as a batch within storage capacity limits. Then the model can find the best time period for bringing in each crude oil batch.

V. REPRESENTING PROCESSES

Process yields need to be included in many models. In many cases, there are different ways that a process can operate. For example, in a refinery there is a process called a reformer, which takes naphtha material, and by heat in the presence of catalyst, convert some molecules in the naphtha to compounds with higher octane or to gases. This results in improved naphtha which is good for making gasoline or for recovery of certain products such as benzene. But the performance of a reformer depends on how it is operated (severity) and the quality of the naphtha used as a feed. To represent this in a model, it is common to include three or four severity modes and allow the LP to interpolate between severity modes to get in-between severities. A mode defines the process as operating in a specific way with a specific feed. Additional vectors (often called delta vectors) are included to adjust the yields from the modes based on one or more key properties in the feed. Fortunately, regarding reformers the yields tend to be convex, which causes interpolation by the LP between adjacent severities. If the LP wants a severity of 96, it would tend to interpolate between 95 and 99 when given choices of 90, 95 and 99 and this approximates reality. When the LP model does not behave in this way, one can use a MIP representation to force choices between adjacent severities or modes. The modern addition to MIP of Special Ordered Sets makes this a very practical choice.

Most of the prior examples have involved planning situations in which we are seeking steady or average operations over a period of time. One might link several time periods together in a single model but each period will be solved to give average results for each. If a process unit is represented by multiple modes, an LP solution might involve several modes at fractional values. This can be interpreted as suggesting operating at a single mode which is a linear interpolation of the modes selected. Or it could be interpreted as operation of each mode for its fraction of the time period. If the latter is a realistic possibility, the LP gives no indication as to when in the period each mode of the operation should take place. This leads us to consideration of scheduling where the timing and sequencing of decisions is important.

VI. A SCHEDULING EXAMPLE

Scheduling is generally beyond the capabilities of LP models and one must use MIP to get proper model behavior. Often this is in discrete time segments but continuous time representation is described in Tahmassebi (chapter 6 in this volume). I will describe an actual model that was developed for an operating plant using MIP.

The situation involved scheduling cracking furnaces. These are large furnaces where heat and pressure crack various hydrocarbon feedstocks to obtain products such as ethylene and propylene. Ethylene and propylene are widely used to make polyethylene and polypropylene films and other materials. There are also many by-products from the cracking process and these range from hydrogen gas to liquids useful in making gasoline and to heavier liquids which are used as fuel. The hydrocarbon feeds are derived mainly from crude oils. The furnaces have tubes through which the feed flows at a constant rate. Burners in the furnace heat the tubes quite hot. As the feed flows through the tubes, it is cracked into different molecules, which are then recovered and separated into the individual products.

The particular cracking furnaces in the application are designed to crack a variety of different feeds although not all furnaces have the same capabilities for handling different feeds or the same capacities or design. Each furnace is considered to have unique characteristics for the model. The ranges of possible feed stocks at the plant are ethane gas, propane gas, butane gas, and naphtha liquid, light gas oil liquid, and heavy gas oil liquid. Each feed type has its own unique cracking characteristics and yields. There are 14 cracking furnaces, each with its own possible feed and feed rate choices. Normally only one type of feed is used at a time although a complication is that some furnaces are designed to handle both a liquid and a small amount of gas feed at the same time; this is called co-cracking.

Cracking operates on a cycle that is different for each feed and each furnace. The cycle starts with feeding the selected feedstock at the selected feed rate and the selected cracking conditions (such as temperature). After a number of days, the cracking will have deposited coke on the inside walls of the tubes and the efficiency of the cracking decreases and there is a danger that a tube might become plugged with coke and cause overheating and possible tube rupture with ensuing fire.

When the feed cracking part of the cycle is at an end, the feed is turned off and a mixture of air and steam is fed in. The oxygen in the air will burn the coke from the inside of the tubes. This is done with the furnace hot. However, every few cycles one need to do more decoking and this is done by cooling the furnace slowly (or else there will be damage) and then doing further work before the furnace can start cracking a new feed. Then once a year, the furnace is taken off line for an extended period to do detailed inspections and repairs and improvements. This is called a turnaround.

The furnaces operate continually around the clock and 7 days per week so at any time each furnace will be in a unique condition. One particular furnace may be on a certain feed and is expected to run 7.33 days before decoking. Its decoking is expected to take 24 hours. Therefore the furnace will be ready for a new feed assignment in 8.33 days. This feed may be different from the previous cycle.

To manage all the individual data we use the OMNI model management system. In this we set up data structures to store data about the individual characteristics of each furnace. This includes what feeds can be used and their minimum and maximum feed rates. We store data for the cracking yields of different feeds at different cracking conditions in each type of furnace. We maintain data on the current status of each furnace. This status includes things like when the current feed cracking started, what is the type of feed (or feeds if co-cracking) and their rates, and what sort of decoking operation will be needed next.

The practical application was to schedule the furnaces, feeds and operating conditions for approximately a month ahead from a designated starting time.

To accomplish the application we need to do considerable data preparation to get starting time conditions each time the application is used. Most of this is performed automatically using our OMNI model management system. The data stored in the OMNI data base is used to project from the starting time for the current cycle on each furnace to the scheduled start time. Then data is projected to the time when the current feed cracking will stop and when decoking will be finished and the furnace will be available for a new feed assignment. A furnace might start the schedule at any stage of a complete cycle from fully committed for the schedule period to available for immediate assignment.

After OMNI has been used to precalculate the starting conditions automatically it is next used to calculate the committed time on each furnace. For operational reasons, a cycle is to be completed with the current feed although changes might be made in feed rate, within the operating limits, or in the heat being applied to cracking. One might shorten the schedule for the cycle on a furnace by 1 or 2 days if that is desirable for the overall schedule but normally the cycle is not made longer because of the dangers of coke blockage in the tubes.

There are limitations on combinations of furnace use because there is a single recovery system for all the furnaces. There is a maximum and minimum limit on the total number of furnaces which are cracking at any one time. There is also a maximum and minimum limit on the number of furnaces cracking particular types of feed. Also, the manpower and facilities required for decoking and for turnarounds is such that only one decoking and one major turnaround can be done at the same time. After decoking, a furnace can be left idle for a few days if necessary to balance the schedule. All of this makes a fairly complex scheduling problem.

Further, there is the question of the objective for the scheduling model. Here we have a situation where another LP has been used to decide what feeds should be cracked to make the required products within the overall plant limits and feed availability and considering the costs and prices. From this LP we have as input to our scheduling MIP model the target amounts of each feed that they would like to crack over the schedule period. We now make our scheduling objective to be the goal-programming values of meeting the target for each feed usage with penalties for being over or under the targets.

With all of the above in mind we build our MIP model. We use OMNI to create a model with a tree structure of decisions. For each furnace we start at the time when the furnace completes the cycle it is already on. We then have as many binary integer variables as there will be feed choices. Each of these will be a particular feed to a particular furnace at a specific time and have a unique time cycle of cracking and decoking. Each binary variable controls a pair of continuous variables, which represent the feed to the furnace at minimum and maximum feed rates. Each binary variable connects to as many further integer variables as there will be future feed choices should MIP select the variable. This continues until no further choices can be made within the span of the schedule. Fortunately the span of the schedule is often one month and some feeds will have a cycle at least that long so the number of integer variables is large but not astronomical. At the start of each choice we include two integer variables representing 1 or 2 idle days on the heater at that time.

For feed usage and furnace limits we use OMNI to calculate the effects on the constraints within the schedule span of time of the cracking operations which were already committed at the start of the schedule and of the new integer variables which might be selected by MIP. For example, to prohibit more than one decoking at the same time, OMNI calculates for each half-day of the schedule period when a decoking would occur and puts a row constraint that only one decoking can occur in that half-day. When an integer variable is on a particular feed, it will have a 1 value in each half-day row involved to limit the number of furnaces on that type of feed. There are many such rows but in addition to modeling a real constraint they tend to limit the solution space and integer nodes so the model solves faster than it would otherwise.

Solution time can be quite long. But our objective of coming close to the target feed rates is a soft objective. By this, I mean that there is little real incentive to get a mathematical optimal if one gets a usable schedule. So typically we only run to a good integer answer and that schedule is acceptable and saves lots of run time on a PC.

This is all set up so that someone without any depth of knowledge about LP or MIP can use the application. We include some penalty vectors, which prevent mathematically infeasible answers but carry a high cost. The OMNI reports point these out to the scheduler as problems that have to be resolved. OMNI is also used before the run to check for a variety of infeasible conditions, and these are reported before the MIP is run.

VII. CONCLUSION

In conclusion, it has been found that when one keeps in mind the behavior of the real system, one can usually build a model that represents the key aspects of it. A tool can also be provided in a form that can be routinely used by people without extensive background in optimization.

Often an LP model will be suitable but in other cases, use of extensions such as MIP or Recursion (Successive LP) will be necessary.

Scheduling in many industries (including refining and petrochemicals where I have the most experience) is becoming of much more interest. Scheduling is inherently more complex than planning, where averages over time are acceptable answers, and scheduling must be concerned with sequencing and timing. Therefore we increasingly find the need to use MIP and other extensions to model the key aspects of scheduling. Fortunately, with these techniques and more powerful computers, it is possible to provide good applications to many more scheduling situations than was once possible.

References

Beale, E.M.L. (1988) *Introduction to Optimization* (New York: Wiley).
Fieldhouse, M. (1993) 'The Pooling Problems', in T.A. Ciriani and R.C. Leachman (eds), *Optimization in Industry* (New York: Wiley).
Haverly, C.A. (1977) 'Behavior of LP Models', *ACM Sigmap Bulletin*, 22:52.

Haverly, C.A. (1978–80) 'Studies of the Behavior of Recursion for the Pooling Problem,' *ACM Sigmap Bulletin*, 25:19 (1978); 26:22 (1979); 28:39 (1980).
OMNI User Manual (May 1997) Haverly Systems Inc.
Tahmassebi, T. (1999) 'A Continuous Time representation for Multi-stage Factory Design and Scheduling', chapter 6 in this volume.

6 A Continuous-time Representation for Multi-stage Factory Design and Scheduling

Turaj Tahmassebi

I. INTRODUCTION

Short-term scheduling of a multi-stage factory consisting of many parallel and competing resources is a problem encountered in many process industries, such as the fast-moving consumer goods', pharmaceutical, food and cosmetics' industries. In such factories, there can be from 2 to 15 packing lines, which can pack one or more formats (sizes). There are several intermediate storage resources as well as several making units. Some packing lines are dedicated to packing different formats of one product, while others can pack different products in different formats. A combination size/product is called Stock Holding Units (SKU). Packing rates depend on both SKU and line. Although it will be assumed throughout that a line packs at its constant nominal rate once started for a SKU, start-up rates are generally slower and a line may take significant time to reach full efficiency. The making units are batch or continuous processing units with finite rates/ capacities. The storage resources store intermediate and finished products. These resources may be dedicated, semi-dedicated or flexible in nature.

In the packing hall, there are two types of sequence-dependent changeovers, minor and major. The former occurs between SKUs of the same size while the latter, which take longer, occur between sizes. Changeover duration may also be packing line-dependent – i.e. may be different for the same two SKUs on different lines. Operators whose number depends on the line and the SKU it is packing operate the lines. The number of operators in the packing room is usually limited, and this limits the number of simultaneously active lines.

Moreover, due to the fact that start-up rates may sometimes be significantly lower than steady-state rates, it is desirable to comply with a minimum run-length constraint before changing over. The storage and making units are also sequence-dependent and their duration depends on the products and resources involved.

For short-term scheduling, usually over a period of 1 week, the objective is to find a schedule with minimum duration, so that high packing/making rates and minimum changeover times are obtained. All SKUs should, ideally, be produced before the end of the horizon. The problem has the features of a resource allocation problem and those of a sequencing problem at the same time. A further complication is introduced by the fact that the lines are interdependent, owing to shared manpower resources and structural constraints. The structural constraints on packing lines indicate whether certain lines are related in operation – e.g. a small group of lines receive materials from a given storage system. These constraints will be treated in this work. There are usually logical constraints on the operation of the storage, making, and packing units. There are limits on the number of storage units that can be accessed from making units, or packing lines.

The main features of the problem are:

- Processing times of activities are not given, but are determined, since the processing of a product item can be shared among several parallel resources.
- Existence of minimum run-length constraints is catered for.
- Existence of common resource constraints (slave resources, manpower, changeover operators), which couple the operation of various parallel resources is modeled. This type of constraint – which is termed *renewable*, since it has to be satisfied all the time – introduces considerable complexity.
- Presence of sequence-dependent changeover, ranging from very small to extremely large duration, found in chemical/consumer good factories are modeled and dealt with, as are constraints on the concurrency of such changeovers across parallel resources.

In the last decade, there has been a growing interest in the design/ scheduling of multi-product batch/continuous plants. Most of the literature in the 1970s and 1980s which considers sequence-dependent set-ups/changeovers deals with the single-line case (Driscoll and Emmons, 1977; Moore, 1975; Robert, 1976). Very few works address the multi-line case. A problem with sequence-dependent set-up costs and nonidentical lines is considered by Prabhakhar (1974) and a mixed-integer formulation along with a branch-and-bound solution procedure is proposed. Geoffrion and Graves (1976) studied a similar problem and proposed a quadratic assignment algorithm to solve it. Parker *et al.* (1977) studied a case with identical lines and sequence dependent set-up costs. In a work by Love and Vegumanti (1978), a model with identical machines and sequence independent set-up costs is considered and they propose a network flows formulation.

In the 1990s, there has been a large amount of interest in the design and short-term scheduling of multi-product batch and continuous plants. A vast number of research articles is available in chemical engineering literature. A review of the relevant research work can be found in Reklaitis (1991). A large number of resources can be scheduled by using the parallel flow-shop problem. The problem objective is usually the assignment of product demands to production lines, the sequencing of product lots on each resource and scheduling of every order (starting and completion times) in such a way that the availability of resources are not exceeded and the customer orders are satisfied in a timely fashion. Musier and Evans (1989) studied the single-stage parallel lines flow-shop problem for process industries. The most significant similarity among different types of problems in industries reported by Musier and Evans was the objective of satisfying demands. However, a particular flow-shop problem that remains unsolved is the problem of satisfying product demands in a batch/continuous multi-product plant with nonidentical parallel lines, sequence-dependent changeovers, due dates for the release of orders as well as constraints on the type of products/variants that can be manufactured/packed on each resource/line. The nature of demands greatly influences the way most multi-product plants are operated. If customer demands are satisfied through stocks and long-term demand forecasts are available, a *campaign operation* can be adopted. In such a case, each production line processes the same sequence of products in cycles having a fixed length of time, batches of the same product are usually produced in sequence (i.e. in a campaign mode). Several scheduling algorithms for single and multi-stage, multi-product plants assuming a cyclic production strategy have been published (Sahinidis and Grossmann, 1991; Voudoris and Grossmann, 1993; Pinto and Grossmann, 1995, 1996). However, in most multi-product plants, like fast-moving consumer goods and chemical specialities, where a wide range of products is manufactured in small tonnage, a campaign operation is not appropriate, as demands cannot effectively be satisfied through product stocks.

A scheduling technique for such manufacturing pattern was proposed by Kondili *et al.* (1993). The work presented a general framework, the State Task Network, for many different scheduling problems arising in multi-product multi-purpose plants. In this framework, the problem is formulated and solved as a mixed-integer linear program (MILP) based on uniform discretization of the time domain. Such Formulations are able to handle different stages, processing of products in parallel identical or nonidentical resources and splitting of materials between stages, limited availability of raw materials and utilities. However, attempting to solve a majority of industrial problems can be difficult as there are usually a large number of integer and continuous variables with many complex constraints in the models with solutions requiring very long CPU times. Gooding *et al.* (1994), who formulated the flow-shop scheduling problem as a travelling-salesman problem and used an enumerative branch-and-bound algorithm represented another approach based on discretization of time. Zhang and Sargent (1994) have proposed a general formulation based on continuous-time representation. The basic idea is to divide the scheduling horizon into time intervals of unknown length. This idea uses events, which coincide with the start or end times of one or more activities. The resulting model is a large-scale mixed-integer nonlinear program, linearized to produce a very large and complex MILP, thereby limiting its application to small problems. Reklaitis and Mockus (1995) developed a similar formulation while Schilling and Pantelides (1996) formulated a simple continuous time version of Resource Task Network.

In terms of possible solution approaches to this problem, Zentner and Reklaitis (1992) presented an approach to reduce the computational cost of the MILP by *a priori* calculation of time events. Pinto and Grossmann (1995, 1996) studied the problem of short-term scheduling for satisfying the demands at a given time horizon. This approach makes use of parallel time axes for resources and activities. Solving realistic industrial problems results in large CPU time requirements. To avoid such problems, the heuristic preordering constraints were used and the resulting mathematical model was solved by decomposition technique.

The formulation presented in this chapter is a continuous-time formulation. At the heart of the formulation is the idea of denoting activities on different resources by a set of events. Start and end times of these events correspond to changes such as the start and end of a changeover. Allocations of events to resources, as well as their relative positions on these resources are determined, such that all resource and structural constraints (constraints relating to the grouping of lines in a multi-stage factory environment affecting the SKU allocation) are satisfied.

This chapter addresses the short-term scheduling problem for multi-purpose/multi-product fast-moving consumer goods' manufacturing plants. A continuous-time representation model for the single-stage packing system, which accommodates sequence-dependent change-overs and minimum run length, is introduced. Different variants of the formulation are examined, together with examples highlighting the differences in computational performance of each variant. This formulation is extended to include making units and multi-purpose storage in a multi-stage plant setting. Scheduling of such plants is complicated by the presence of shared resources – which link the operation of the lines, and structural constraints linking the operation of different stages.

Computational experiments using standard solvers such as XPRESS-MP and CPLEX suggest that this continuous-time approach has the potential to outperform other methods, because of the simplicity of the logical decisions involved and the efficiency of the formulation. Preliminary results on the solution of a real industrial problem will be presented. These show the considerable promise of the approach for solving real, large-scale scheduling problem in the process industry.

II. PROBLEM DEFINITION

The short-term scheduling problem for multi-product and multi-resource and multi-stage factories involving many parallel lines to be tackled in this work has the following features:

- Each SKU demand can be packed on a set of resources. Multiple lots of the same SKU/product are allowed. The same SKU/product may be packed/processed on several lines in the same schedule
- The demand for each SKU/product is specified. The SKU/product demands have to be satisfied at the due dates or by the end of the scheduling horizon
- Each SKU/product can be packed/processed on a sub-set of the available lines
- The packing/making time for a SKU depends on both the nature of the SKU/product and the type of the resource/line. The minimum

run time depends on the SKU/product and the type of packing/making resource

- Before starting a new activity, a changeover for cleaning and resource set-up is needed. The changeover time is sequence-dependent on the nature of the SKUs/products and the sequence.

Because of flavor and/or color incompatibilities, some SKU/products sequences may not be allowed. Resource constraints relating to raw materials or limited manpower or concurrent changeover limitations have to be considered.

Given all these features, the scheduling problem consists of determining, the allocation of SKUs/products to lines, and the SKU sequence on each line and the scheduling timetable so that all scheduling constraints are taken into account while optimizing an objective (e.g. makespan).

III. MATHEMATICAL MODEL

A mathematical formulation for the parallel-resource, multi-product packing-hall scheduling problem will be developed in this section. First, a basic formulation is developed for the scheduling of the nonidentical parallel lines problem. Three versions of the formulation will be presented. The formulations of the constraints for the common resources and the concurrency of parallel activities are developed in the next section. The main features of the proposed model describe below are: (1) The use of continuous-time representation, and (2) the decision variables corresponding to the allocation of SKUs to lines and the relative positions of time activities on the resources. As will be shown, the formulation given here can easily account for sequence-dependent changeovers and some other real constraints found in industry.

3.1. The objective function

The objective of minimizing the makespan is equivalent to minimizing the maximum completion time of all the activities – i.e. min η = min max (end time of all events). Using a standard linear programming formulation, this can be expressed as in (6.1), where η is defined as the schedule completion time.

Alternative scheduling criteria can be chosen, production schedule makespan, schedule tardiness, making to stock, etc. The objective selected here is the packing schedule makespan

$$\min \eta \qquad (6.1)$$

3.2. Problem constraints

The description of the variables, parameters, and the constraints in the model will be covered in the following sections.

- **Definition of start and completion times for SKUs/products p.** Define the nonnegative continuous variables, S_{pl} as the start time of SKU/product p on line l, E_{pl} as the end time of SKU/product p on line l.

- **Every SKU/product p which can be packed/made on a line l can be allocated to that line for a minimum run time greater or equal to zero.** In commercial applications, the packing or production times of all SKUs/products are limited by a minimum value (e.g. the minimum run time is limited by a half-shift, 4 hours). In this chapter, the minimum value has an additional meaning. The minimum run time is a threshold below which all events will be treated as zero duration events, such events will be placed at the origin. Define Γ_l as the set of SKUs/products that can be packed on line l, m_{pl} as the minimum run time of item p on line l. Define the binary parameter γ_{pl} which has a value of 1 if the SKU/product is allocated to resource l, otherwise zero. Also, define the following conditions for the size of the events active on the lines. The duration of an event is either zero or greater than the minimum run time m_{pl}

$$E_{pl} \leq M \gamma_{pl} \qquad \forall l, p \in \Gamma_l \qquad (6.2)$$

$$E_{pl} - S_{pl} \geq m_{pl} \gamma_{pl} \quad \forall l, p \in \Gamma_l \qquad (6.3)$$

M is assumed to be the available time in the scheduling problem (usually around 120 hours).

- **Every SKU/product p which can be packed/made on a line l is positioned either before or after SKU q which can also be packed/made on the same line.** The terms before and after are interpreted as some time before or after. The following relationships define the sequence dependency between the events. These relationships are also known as the *classical disjunctive constraints* for the basic scheduling problem.

Let p and q index SKUs/products and l index lines. Also, define the binary variables δ_{pql} which is defined with a value of 1 if the occurrence of event p is before that of q on line l and value of zero if p is after q on line l. The variable δ_{pql} has been defined for the case where the subscript p is less than q. As the variables δ_{pql} and δ_{qpl} sum to unity, those variables of δ for the case of q greater than p have been eliminated. This reduces the number of binary variables considerably. Also, define Δ_{pql} as the changeover time from p to q on line l.

$$S_{ql} \geq E_{pl} - M\delta_{qpl} + \Delta_{pql}\gamma_{pl}, \qquad \forall l \text{ and } p,q \in \Gamma_l \mid p > q \qquad (6.4)$$

$$S_{pl} \geq E_{ql} + M(\delta_{qpl} - 1) + \Delta_{qpl}\,\gamma_{ql} \quad \forall l \text{ and } p, q \in \Gamma_l \mid p < q \qquad (6.5)$$

M has been defined similar to that in section above.

- **Makespan definition** (η). Let η be the time needed to complete the production requirements of the scheduling problem (i.e. the makespan). The definition of η should be included as a problem restriction when the minimization of the makespan is chosen as the problem objective. The value of η is determined as the largest completion time in the scheduling problem. The completion time of all events must be such that they are bounded by the makespan variable η as given by (6.6),

$$\eta \geq E_{pl} \; \forall l, p \in \Gamma_l \qquad (6.6)$$

- **Demand constraints.** Define r_{pl} as the rate of packing item p on line l which is equal to zero if item p cannot be packed on line l. The formal constraints describing the scheduling problem formulation are as follows

$$\sum_l r_{pl}(E_{pl} - S_{pl}) = D_p \quad \forall p \qquad (6.7)$$

The various components of the model act as follows. The objective function (6.1), along with constraints (6.6) lead to the minimization of the maximum end time of all items on all lines, which is equivalent to minimizing the makespan. The explanation for constraint (6.4) is as follows. When δ_{qpl} is equal to 0 and q occurs after p on line l, constraint (6.4) is active, ensuring that the start time of q is greater than the end time of p by at least an amount equal to the necessary changeover time Δ_{pql}. It must be pointed out that p is greater than q in (6.4). (6.5) behaves very similarly to (6.4), except that (6.5) is valid for the case where p is less than q.

Constraints (6.2) and (6.3) ensure that when a SKU/product is not assigned to a line (i.e. $\gamma_{pl} = 0$), then its end time on that line is equal to zero, and consequently its start time; while if $\gamma_{pl} = 1$, the duration of events is greater than the threshold parameter. It is worth noting that when $\gamma_{pl} = 0$, there is no changeover between SKU/product p and other SKUs/products on line l. This is ensured in constraints (6.4) and (6.5) by multiplying the changeover time with the appropriate γ_{pl} and γ_{ql} in the terms $\Delta_{pql}\gamma_{pl}$ and $\Delta_{qpl}\gamma_{ql}$. Constraints (6.7) ensure that the demand for each item is met.

All the constraints (6.2)–(6.7) as well as the objective function for the problem as defined by (6.1) define the basic formulation known as the classical disjunctive problem, given here for clarity as formulation (A). As formulation (A) contains the parameter M, it may have poor lower-bound characteristics for some commercial problems. This results in the objective values of the integer solutions being very different from the relaxed/lower bound values. It is possible to tighten the formulation (A) by introducing additional constraints. In the formulation (A), it is implied that,

$$\gamma_{pql} + \gamma_{qpl} = 1 \quad \forall l \text{ and } p, q \in \Gamma_l \mid p < q \tag{6.8}$$

To tighten the formulation, an alternative definition for δ is introduced. The binary variable δ_{pql} is defined to be 1 if p and q are packed on line l and p is packed before q and both p and q are packed on line l, otherwise zero. To enforce the above definition, the following constraints in the basic formulation are introduced,

$$\delta_{pql} + \delta_{qpl} = \gamma_{pl}\gamma_{ql} \quad \forall l \text{ and } p, q \in \Gamma_l \mid p < q \tag{6.9}$$

To implement (6.9), this equation needs to be linearized as follows. Values of $\gamma = 0$ cause the summation term in (6.9) to be equated to zero. This implies that the values of δ become dynamically dependent on γ and the summation term for δ can be equal to 1, iff both values of γ for p, q are equal to 1. Using a standard derivation, the following relationships may be defined

$$\delta_{pql} + \delta_{qpl} \leq \gamma_{pl} \quad \forall l \text{ and } p, q \in \Gamma_l \mid p < q \tag{6.10}$$

$$\delta_{pql} + \delta_{qpl} \leq \gamma_{ql} \quad \forall l \text{ and } p, q \in \Gamma_l \mid p < q \tag{6.11}$$

$$\delta_{pql} + \delta_{qpl} \geq \gamma_{pl} + \gamma_{ql} - 1 \quad \forall l \text{ and } p, q \in \Gamma_l \mid p < q \tag{6.12}$$

(6.4) and (6.5) in the basic formulation (A) can be combined in the form of

$$S_{ql} \geq E_{pl} + M(\delta_{pql} - \gamma_{pl}) + \Delta_{pql}\gamma_{pl} \quad \forall l \text{ and } p, q \in \Gamma_l \mid p \neq q \tag{6.13}$$

Adding (6.9), (6.10), (6.11), (6.12), and (6.13) to the basic formulation (A) results in a tightened formulation. This formulation, referred to as the *classical ordering* formulation, is defined as formulation (B). It must also be pointed out that all binary variables δ are present in formulation (B) and a sub-set cannot be deleted as in formulation (A).

IV. MODIFIED PROBLEM REPRESENTATION USING VEHICLE ROUTING PROBLEM (VRP) FORMULATION

To provide a more thorough understanding of the sequencing problem, an alternative formulation was proposed. This formulation has a major difference from formulations (A) and (B), the sequencing variables are defined with respect to the *immediate precedence of all events active on lines*. Using the notation described above it is possible to present an alternative formulation. Introduce the following variables

$$z_{pql} = 1, \quad \text{if } p \text{ is produced directly before } q \text{ on line } l$$
$$0, \quad \text{otherwise} \quad \forall l \text{ and } p, q \in \Gamma_l \mid p \neq q \tag{6.14}$$
$$z_{pl}^{beg} = 1, \quad \text{if line } l \text{ starts with product } p$$
$$0, \quad \text{otherwise} \quad \forall l \text{ and } p \in \Gamma_l \tag{6.15}$$
$$z_{pl}^{end} = 1 \quad \text{if line } l \text{ ends with product } p$$
$$0, \quad \text{otherwise} \quad \forall l \text{ and } p \in \Gamma_l \tag{6.16}$$

The variables z_{pql} define the position of all SKU/products events on lines l relative to each other. The remaining z^{beg} and z^{end} variables define the first and last SKU/products events on all lines. Define the constraints for the Vehicle Routing Problem formulation as defined for the scheduling problem

$$\sum z_{pl}^{beg} \leq 1 \quad \forall l \tag{6.17}$$

$$\sum z_{pl}^{end} \leq 1 \quad \forall l \tag{6.18}$$

(6.17) and (6.18) show that sequencing SKUs/products on line l results in only one SKU/product at the start of the sequence and only one at the end point. In the case of only one SKU/product present on a line, the same SKU/product can be at the beginning and the end at the same time. To describe the immediate precedence relationship between the SKUs/products, some new relationships are required. The following relationships describe the timings of SKUs/products directly before p

$$z_{pl}^{beg} + \sum_{\substack{p \in \Gamma_l \quad q \in \Gamma_l \\ p \neq q}} z_{qpl} = \gamma_{pl} \quad \forall l \text{ and } p \in \Gamma_l \tag{6.19}$$

(6.19) indicates that conditional on SKU/product p existing on line l, there can be only one other SKU/product q which precedes it. If SKU/product p is not the first event, there must be a SKU/product q, which occurs before p

$$z_{pl}^{beg} + \sum_{p \in \Gamma_l \quad q \in \Gamma_l} z_{pql} = \gamma_{pl} \quad \forall l \text{ and } p \in \Gamma_l \tag{6.20}$$

(6.20) indicates that conditional on SKU/product p existing on line l, there can be only one SKU/product q, which precedes it. If p is not the end SKU/product, then there must be an additional SKU/product q which occurs after p. The positions of all SKU/product events p are subject to other operating constraints – i.e. the duration of changeovers on lines. The nonoverlapping of SKU/product events on the lines and the existence of the sequence-dependent changeovers on lines can be enforced as follows

$$S_{ql} \geq E_{pl} + M(z_{pql} - \gamma_{pl}) + \Delta_{pql} z_{pql} \quad \forall l \text{ and } p, q \in \Gamma_l \mid p \neq q \tag{6.21}$$

(6.21) implies that if $z_{pql} = 1$ then it follows that γ_{pl} cannot have a 0 value, the value of 0 for γ results in an infeasible solution. This implies that the immediate precedence between SKUs/products p and q can occur only if the SKU/products p can be packed on that line. In this case, only a value of $\gamma_{pl} = 1$ is a valid solution to the problem. If $z_{pql} = 0$ then γ_{pl} can be either 0 or 1. The value of 0 for this variable may be interpreted as either the SKU p is not directly before q or one or both SKUs do not exist on the active line.

Also, introduce a new relationship, a global time bound as

$$\sum_{p \in \Gamma_l} (E_{pl} - S_{pl}) + \sum_{\substack{p,q \in \Gamma_l \\ p \neq q}} \Delta_{pql} \, z_{pql} \leq \eta_l \quad \forall l \qquad (6.22)$$

The first term in (6.22) refers to the packing/making times on a given line. The second term refers to all changeover times active on line. The relationships (6.1), (6.2), (6.3), (6.6), (6.7) together with the relationships for Vehicle Routing Problem formulation (6.17)–(6.22) define the alternative formulation known here as formulation (C).

At this stage it is useful to compare this formulation with other continuous-time counterparts. Works such as Zhang and Sargent (1994), Schilling and Pantelides (1996), and Pinto and Grossmann (1995, 1996) define the binary variables dependent on a multiple of tasks/activities and the number of variable time-slots in the problem. The number of the slots is usually made equal to the number of activities. The binary variables would be equivalent to a number greater than $\sim 2(p)^2 l$. The number of continuous variables will be proportional to the $\sim (pl)^3$. The corresponding numbers for this formulation are $\sim (p)^2 l$ and $\sim 2(pl)$, respectively.

V. COMMON RESOURCE CONSTRAINTS

5.1. Parallel events constraints

To model the concurrency of parallel events, note that parallel resource consumption increases only when the activity starts. Therefore, it is sufficient to enforce the parallel resource constraints only at these instances. To this end, we define overlap in the following way. An item p on line l overlaps item p' on line l' if the activity denoted by pl starts while the packing activity denoted by $p'l'$ is ongoing. It is worth emphasizing that, in this sense, overlap is not only nonsymmetric, but mutually exclusive, in the sense that if pl overlaps $p'l'$, it cannot be that $p'l'$ overlaps pl. Note also that for overlap l must be different from l' while p may be equal to p'.

According to the above definition, packing activity pl overlaps packing activity $p'l'$ iff the start of pl is greater than or equal to the start of $p'l'$ and simultaneously less than or equal the end of $p'l'$. Define the binary variables $\alpha^{(1)}_{plp'l'}$, $\alpha^{(2)}_{plp'l'}$ and $\alpha^{(3)}_{plp'l'}$. The first variable $\alpha^{(1)}$ has a value of 1 if packing activity pl overlaps packing activity $p'l'$, otherwise zero. The variable $\alpha^{(2)}$ has a value of 1 if the start of packing activity pl is strictly less than that of packing activity $p'l'$ otherwise zero. The variable $\alpha^{(3)}$ has a value of 1 if the start of packing activity pl is greater than or equal to the end of packing activity $p'l'$ otherwise zero.

Therefore, the relationship between various events are represented by the following constraints

$$S_{pl} - S_{p'l'} \leq -M\alpha^{(2)}_{plp'l'} + M + \varepsilon \qquad \forall\, p,l,p',l' \,|\, l \neq l' \tag{6.23}$$

$$E_{p'l'} - S_{pl} \geq M\alpha^{(1)}_{plp'l'} - M + \varepsilon \qquad \forall\, p,l,p',l' \,|\, l \neq l' \tag{6.24}$$

$$S_{pl} - S_{p'l'} \geq M\alpha^{(1)}_{plp'l'} - M \qquad \forall\, p,l,p',l' \,|\, l \neq l' \tag{6.25}$$

$$S_{pl} - E_{p'l'} \geq M\alpha^{(3)}_{plp'l'} - M \qquad \forall\, p,l,p',l' \,|\, l \neq l' \tag{6.26}$$

(6.24) and (6.25) indicate the overlapping of the activities of pl and $p'l'$. (6.23) and (6.26) show the nonoverlapping between these two activities. The binary variables α should satisfy the relationship given below

$$\alpha^{(3)}_{plp'l'} + \alpha^{(2)}_{plp'l'} + \alpha^{(1)}_{plp'l'} \qquad \forall\, p,l,p'l' \,|\, l \neq l' \tag{6.27}$$

The variables $\alpha^{(1)}$, $\alpha^{(2)}$, and $\alpha^{(3)}$ are defined in order to represent the overlapping or nonoverlapping of the events adequately. All such variables are necessary for the following reason. If only one variable α is defined in the above formulation, there is only one value of $\alpha = 0$ that will satisfy (6.23)–(6.26). This will result in all these relationships becoming redundant and therefore the overlapping events will not be captured correctly. The value of $\alpha = 1.0$ will make the above relationship infeasible and therefore unacceptable. Defining only two variables $\alpha^{(1)}$, $\alpha^{(2)}$ to represent overlapping and nonoverlapping will not be adequate as values of these binary variables will be selected by the MILP in such a way that will indicate nonoverlapping of activities, even when there are true overlapping of activities taking place. Now to enforce the requirement that manpower consumption is always less than a given upper limit, it is sufficient to add the following set of constraints

$$\lambda_{pl} + \sum_{p'} \sum_{l' \neq l} \lambda_{p'l'} \alpha^{(1)}_{plp'l'} \leq L \qquad \forall\, l, p \in \Gamma_l \tag{6.28}$$

The constant λ_{pl} denotes the absorption coefficient (inverse rate) for the activity pl.

5.2. Parallel changeover constraints

The changeovers are usually in two classes, minor and major. Minor changeovers are of short duration, usually consist of cleaning operations and are produced by either equipment or operator. In contrast, major changeovers are lengthy and usually require skilled operators to perform them. The major changeovers can occur at the same time on different resources (concurrent changeovers). If the concurrency of the changeovers is problematic, then changeover events occurring at the same time need to be avoided. Usually, concurrency of minor changeovers does not cause any problems, changeover interactions can be assumed to cause problems only as far as the 'major changeovers' are concerned.

For mathematical simplicity, make the following assumptions. Define u to be a major changeover with the start time and end time S_u and E_u, respectively. Also define v to be a major changeover with the start time and end time S_v and E_v. These start and end times are a mapping of other start and end times occurring on parallel lines. Let this mapping be defined by

$$S_u = E_{pl} \quad E_u = S_{ql} \quad S_v = E_{p'l'} \quad E_v = S_{q'l'} \quad \forall u, v \, u \neq v \qquad (6.29)$$

where p and q activities occur on line l in such a way that activity q follows p after a major changeover. Also activities p' and q' occur on line l' in such a way that q' follows p' after a major changeover. In order to establish whether the changeovers are overlapping, the existence of the changeovers u must be established. Now define the variable γ_u with a value of 1 if event u exists and 0 otherwise, which establishes the following relationship

$$E_u - S_u \geq \varepsilon \gamma_u \quad \forall u \qquad (6.30)$$

$$E_u - S_u \leq M \gamma_u \quad \forall u \qquad (6.31)$$

Also, define the variables θ_{uv} with a value of 1, if major changeovers u and v occur and u is before v and 0, otherwise. In order to avoid the concurrency of the changeovers, formal constraints need to be introduced. The form of the constraints will be as follows

$$E_u - S_v \geq \Delta_u + \Delta_v + \theta_{uv} M + (\gamma_u - 1)M + (\gamma_v - l)M \qquad (6.32)$$
$$\forall u, v \, u \neq v$$

$$E_v - S_u \geq \Delta_u + \Delta_v + (\theta_{uv} - 1)M + (\gamma_u - 1)M + (\gamma_v - 1)M \qquad (6.33)$$
$$\forall u, v \, u \neq v$$

where Δ_u and Δ_v are the appropriate changeovers. Clearly, for the two major changeovers not to overlap, one has to succeed the other, which in turn means that the period from the beginning of one nonpacking event on one line to the end of the packing event on the other must be equal to or longer than the sum of the two changeover periods. The constraints (6.32) and (6.33) ensure this. The value of 1 for θ_{uv} will make the relationship (6.33) active and make (6.32) redundant.

It is also worth noting that enforcing the nonconcurrency of major changeovers in this way is valid for possible limiting conditions, such as the case when $E_v - S_u = 0$ or the case when $S_u = 0$ and $E_u \neq 0$. If $S_v = E_v = 0$, then the difference is of zero length and will set γ_v to 0 and hence will render the constraint redundant. Only values of both γ_u and γ_v to set 1 will make the constraints active.

It can be proved that the inclusion of the existing binary variables in the formulation of the concurrency problem is not necessary and sufficient to determine the existence of a major changeover on a line. Other binary variables and additional constraints are necessary to adequately determine the position of the major changeovers and the concurrency between them. Since, the implementation of an accurate formulation requires a large number of additional variables and constraints to the basic formulation (A or B) an alternative formulation as an extension to formulation (C) was developed. In this section, the nonconcurrency of changeover constraints for the vehicle routing problem will be given.

Define the index u as the index of major changeovers defined by (p_u, q_u, l, Δ_u). p and q are events and which the changeover u occurs between them with duration Δ_u. Also, define the following variables, μ_{uv} with a value of 1, if the major changeovers u and v occur and u is done before v and 0 otherwise. The definition of μ is dependent on the existence of the major changeovers. For this purpose, define binary variables, z_{ul} as the variable denoting the existence of u on l. Also, define s_u as the start time of changeover u $\forall u, v\, u \neq v$. The schematic diagram of the major changeover u and v is given in Figure 6.1. Additional constraints are necessary to define the existence and ordering of changeovers.

$$\mu_{uv} + \mu_{vu} = z_{ul}\, z_{vl} \quad \forall u, v\, u \neq v \text{ and } l \neq l' \tag{6.34}$$

To implement (6.34) in a linear programming framework, a linearization is required. Values of 0 for either z_{ul} or $z_{vl'}$ cause the summation term in (6.32) to be made equal to 0, indicating that the variables μ are no longer required.

The major changeovers usually occur between the activity events in such a way that there are idle times present as well as the duration of changeovers. This means that in order to define the start time of the

Figure 6.1 Schematic representation of events, major changeovers, and start and end times

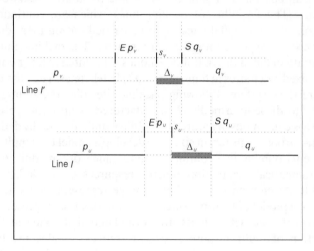

changeovers, additional constraints are needed. These constraints relating to the timings of the changeovers are outlined by the following relationships. E_{ul}^p is defined as the start time of the time gap between the end time of event p and the start time of event q, and S_{ul}^q is defined as the end time of the gap between the end time of event p and the start time of event q. γ_{ul}^p is defined as the variable denoting the existence of the event p_u on line l.

$$s_u \geq E_{ul}^p - M(\gamma_{ul}^p - z_{ul}) \qquad \forall u \qquad (6.35)$$

$$s_u \leq S_{ul}^q - \Delta_{ul} \, \gamma_{ul}^q + M(\gamma_{ul}^q - z_{ul}) \qquad \forall u \qquad (6.36)$$

The nonoverlapping constraints can be written as,

$$s_v \geq s_u + \Delta_{ul} \, z_{ul} - M(z_{ul} - \mu_{uv}) \qquad \forall u,v, u \neq v \text{ and } l \neq l' \qquad (6.37)$$

The start time of the changeover v is positioned after the changeover u by at least an amount equal to the changeover Δ_u.

VI. MULTI-STAGE MODEL FORMULATION USING CONTINUOUS-TIME PRESENTATION

The consumer goods' manufacturing scheduling problem is multi-stage in nature consisting of several active stages such as making, packing, and post-dosing units often separated by multi-purpose storage resources of finite capacities. The packing and making resources are usually dedicated, semi-dedicated or sometimes flexible in nature, of finite rates/capacities. The flexibility of packing lines, which are continuous in operation, allows parallel packing to be carried out on them drawing similar product types from upstream resources. The making units usually operate in batch or continuous modes. The batch making units are making fixed batch sizes in finite processing batch times. The post-dosing units are slave systems following packing rates directly, adding other products to those intermediates manufactured by upstream resources and always continuous in nature. The storage units are usually dedicated or multi-purpose of different sizes. To develop a model for a multi-stage factory environment in a continuous-time framework, modifications to the mathematical formulation of active resources are needed. There is a need to develop an intermediate storage representation within the multi-stage model of factory using the continuous-time representation. In this work, the active units are formulated first followed by the description of multi-purpose storage resources. Finally, the linking between various resources is modeled to achieve consistent timing between all activities in the multi-stage model.

6.1. Modifications to making/packing activities

The formulation of the making and packing resources in a multi-stage framework is similar in nature to the parallel resource formulation described in earlier sections. The additional complexity is the existence of the sub-lots for a SKU/product/line allocation. All variables defined in the previous sections should be used with an additional superscript defining the sub-lot on an event/activity. The resulting variable S_{pl}^{n} defines the sub-lot of the event pl. All relationships defined above are valid in the multi-stage framework. The Vehicle Routing Formulation for a single-stage problem can easily be modified to cater for the case where each lot size has several sub-lots. In the following sections, however, the subscripts for sub-lots are dropped from the relationships for convenience. These sub-lots have been fully utilized in the model implementation phase.

6.2. Intermediate storage formulation

In order to develop this formulation, the following considerations are made. All events acting on an intermediate storage l have been classified as follows:

events : {*inputs*}, {*outputs*}

For each event, define two event points E and S. Considering the dynamics of the product activities/events active on the storage resources, each input and output product sequence should be the same – e.g. if a product enters an intermediate storage resource, it is unlikely that another product leaves that intermediate resource. This leads to the definition of envelopes of input and output product events active on the intermediate storage resources. An envelop with an input and output event active on a tank defines a case where two simultaneous input/output events can be active on the resource. In this case, only one downstream resource is drawing required products from that resource and only one upstream resource is feeding that same resource. In the case of a three-stage system of making–storage–packing, an envelop with one input and one output product events defines a situation where only one packing event is drawing product from a storage resource and only one making unit feeds the product into the storage. An envelop of one input and two output events defines a case where one input and two output events are taking place on a storage unit. In such case, two packing events may be overlapping and drawing a product from a storage resource and only one input is feeding that resource. To represent the product sequencing for input and output events, the Vehicle Routing Formulation is used in order to sequence the envelopes acting on the intermediate resources. This part of the theory is similar to the basic formulation and has therefore been omitted.

Define variables $\delta_{ii'\xi l}$ where i' exists after i within the envelop ξ on a resource l. Also, define variables $\delta_{i\xi l}^{end}$ where i belongs to the set of end times and $\delta_{i\xi l}^{beg}$ where i belongs to the set of start times. Define, $\tau_{i\xi l}$ as continuous variables mapping all events (start and end times) acting on l within ξ. Define variables $x_{ii'\xi l}$ as the relative timing between product events i acting on l within ξ. Also, define variables $w_{i\xi l}$ as the stock of product events i acting on a resource l.

The ordering of events i may be arranged by the *classical ordering* formulation. This should also take into account the zero length or otherwise of the input/output events. The following constraints are written for the mapping variables τ

$$\tau_{i'\xi l} \geq \tau_{i\xi l} + M\delta_{i\xi l}^{beg} - M \quad \forall \, i\xi l \text{ and } i \neq i' \tag{6.38}$$

$$\tau_{i'\xi l} \geq \tau_{i\xi l} - M\delta_{i\xi l}^{end} + M \quad \forall \, i\xi l \text{ and } i \neq i' \tag{6.39}$$

$$\tau_{i'\xi l} \geq \tau_{i\xi l} + M\delta_{i'i\xi l} - M \quad \forall \, i\xi l \text{ and } i \neq i' \tag{6.40}$$

The stock updating equations within the envelop of events is as follows,

$$y_{\xi l}^{close} = y_{\xi l}^{open} + w_{i\xi l} \quad \text{if} \quad \delta_{i\xi l}^{end} = 1 \tag{6.41}$$

$y_{\xi l}^{close}$ and $y_{\xi l}^{open}$ are the closing and opening stocks for the envelop ξ. The variable $x_{ii'l}$ is evaluated as below. Note that r refers to the input rate in (6.42) and refers to the output rate in (6.43).

$$x_{ii'\xi l} = (\tau_{i\xi l} - \tau_{i'\xi l})r_{i\xi l} \quad \forall \ i\xi l \ \text{and} \ i \neq i' \tag{6.42}$$

$i \in$ input, start, and end time events   if $\delta_{ii'\xi l} = 1$

$$x_{ii'\xi l} = (\tau_{i\xi l} - \tau_{i'\xi l})r_{i\xi l} \quad \forall \ i\xi l \ \text{and} \ i \neq i' \tag{6.43}$$

$i \in$ output, start, and end time events if $\delta_{ii'\xi l} = 1$
The variables $w_{i\xi l}$ are calculate as follows

$$w_{i\xi l} = \sum_{\substack{i \in \text{input} \\ i \in \text{start time} \\ i \neq i}} x_{ii'\xi l} - \sum_{\substack{i \in \text{input} \\ i \in \text{end time} \\ i \neq i}} x_{ii'\xi l} - \sum_{\substack{i \in \text{output} \\ i \in \text{start time} \\ i \neq i}} x_{ii'\xi l} + \sum_{\substack{i \in \text{output} \\ i \in \text{end time} \\ i \neq i}} x_{ii'\xi l} \tag{6.44}$$

$y_{\xi l}^{close}$ and $y_{\xi l}^{open}$ should lie within 0 and maximum capacity limits as defined by the problem. These variables should follow the relationship, which is given below. For this relationship to be true the envelopes ξ and ξ' must be adjacent.

$$y_{\xi l}^{close} = y_{\xi' l}^{open} \quad \text{if} \quad z_{\xi \xi' l} = 1 \tag{6.45}$$

6.3. Logical linking constrains

Additional constraints

These are constraints which in certain cases may not be crucial, but their inclusion in the problem formulation frequently accelerates convergence to the optimal solution or provides meaningful results. Alternatively these constraints are required in order to generate practical solutions to the problem in hand. One such constraint of the first type is the time constraints imposed on the completion time of a schedule, given that a feasible schedule has already been derived. An example of the second type of constraint is the linking constraint in a multi-stage factory model. In this work, the latter constraints are explained in detail.

To achieve linking between stages, the events relating to the lots/sub-lost as well as the flow of materials/products need to be linked up. For clarity, the notion of a product/variant sub-lot is introduced here. Each lot from a product/variant family may consist of two or more sub-lots.

To give an example of such constraints, define the variable n_R^ξ as the sub-lot n of the variant/product of ξ, a generic product/base on the resource R, where R is defined to be a resource such as making, storage, slave utility or a packing system resource. Furthermore, define the following sets denoted by N, Θ, H, and Γ_R for describing the relationship between variables defined below. Define N as

$$\{n_R^\xi, n_{R'}^{\xi'}\} : N$$

Where the lot sizes n_R^ξ and $n_{R'}^{\xi'}$ on resource R and R' are related to each other. If this is true, then the set N carries an entry of unity for that pair, otherwise zero.

Define the set Θ as given by

$$\{R, R'\} : \Theta$$

If resources R and R' are connected, then the corresponding element of the set Θ will be 1, otherwise 0.

Define H as

$$\{\xi^R, \xi^{R'}\} : H$$

H implies that the variant ξ on R originates from ξ' on R' – e.g. the product p on packing line l originates from variant v on slave resource s. Finally, Γ_R defines the set of products, which can be allocated to resource R.

In this formulation, the events in two adjacent stages are linked together logically depending on the physical structural connections and the possibility of some events being nonzero. It is anticipated that, although all active resources in stages are usually connected on a static basis, on a dynamic basis, the resources may not always be connected.

$$\begin{aligned}
\text{Either}: & \quad \{E(n_R^\xi) - E(n_{R'}^{\xi'}) = 0 \quad \text{And} \quad S(n_R^\xi) - S(n_{R'}^{\xi'}) = 0\} & (6.46)\\
\text{Or}: & \quad \{E(n_R^\xi) = 0 \quad \text{And} \quad S(n_R^\xi) = 0\} \quad \text{for all } n_R^\xi, n_{R'}^{\xi'} \in N\\
& \hspace{9cm} \text{for all } R, R' \in \Theta\\
& \hspace{9cm} \text{for all } \xi_R, \xi'_R \in H
\end{aligned}$$

VII. COMPUTATIONAL EXPERIENCE AND CASE STUDIES

7.1. Example 1

Small-scale packing-hall scheduling problem

Consider a scheduling problem involving 6 packing lines in fast-moving consumer goods' manufacturing plant. Assume that there are 9 products to be packed in this factory. The matrix representing the packing rates for different packing lines are given in Table (6.1). The products are subject to weekly demands as given by Table (6.2).

In this example it is assumed that there are sequence dependencies between SKUs resulting in different changeover times on various lines. Table 6.3 gives the changeover data. The available packing time was assumed to be 120 hours.

The mixed integer model of formulation (A) was implemented on a Pentium Pro 200 computer using XPRESS-MP mathematical programming software. The resulting model has 190 decision variables of which 69 are binary variables. The total number of constraints in the model is derived to be 179. The objective function was assumed to be the minimization of a unified completion time for all packing lines. The value of M was assumed to be 120 hours. This acts as an upper bound on the end

Table 6.1   Packing rates (t/h) of various SKUs on different packing lines

SKU/line	1	2	3	4	5	6	7	8	9
L1	2.0	–	–	–	1.0	1.0	2.0	1.0	–
L2	–	–	–	1.0	–	1.0	–	2.0	1.0
L3	–	1.0	–	–	–	–	1.0	–	–
L4	1.0	–	1.0	–	1.0	–	–	1.0	–
L5	1.0	1.0	1.0	1.0	–	–	–	1.0	1.0
L6	–	–	1.0	–	1.0	–	1.0	–	1.0

Table 6.2   Products demands (t)

SKU	1	2	3	4	5	6	7	8	9
	18	10	100	100	10	10	100	10	30

time events in the model. It must be pointed out that if a formulation based on a uniform discretization interval had been chosen, the number of binary decisions for this problem would have been ~12 960. The corresponding continuous time formulations reported in the literature would have the number of binary and continuous variables in excess of 1250 and 6000 respectively, mainly due to the 0.5 hour changeover data resulting in a discretization interval of 0.5 hour. A relaxed solution of 62.53 was obtained in 81 iterations of the simplex. A first-integer solution of 68.13 was obtained by exploring 50 of the nodes in the branch-and-bound tree. The second solution of 64.57 was obtained after 59 nodes of the tree. The third and optimal solution of 63.02 was obtained after 65 nodes. This solution was obtained in around 0.6 seconds CPU. The Gantt chart of the solution has been shown schematically in Figure 6.2. The reason for optimality of the solution is explained below. The tightened formulation (B) was set up for this packing hall problem. This model has 164 variables of which 113 are binary variables. There are 311 constraints with 1099 nonzero elements. A relaxed solution of 62.53 was obtained. The same solution of 63.02 was obtained in 60 nodes, in 0.55 seconds CPU time. The formulation of Vehicle Routing Problem as defined by formulation (C) was set up for this problem. There are 214 variables of which 163 are binary variables. There are 241 constraints with 1066 binary nonzero elements. A relaxed solution of 62.66 was obtained. The solution of 63.02 was obtained in 0.1 seconds CPU in 21 nodes of the branch-and-bound tree. No other solution was generated. The solution can be proved to be optimal. The search was terminated automatically in 6078 nodes of the tree. The termination was achieved in 56 seconds CPU. It must be pointed out that although formulations (A) and (B) managed to get the right solution, the automatic termination by the solver could not be achieved.

The optimized solution indicates that products p3, p4, p5, p8, and p9 have been split on different packing lines. Line 5 has been dedicated to product 4, which constitutes the line with the longest completion time. Also line 4 has been dedicated to product 3. The allocation of SKUs to other lines is such that the completion times are generally unconstrained but equal or less than that of lines 5 and 4. This corresponds to the optimal solution. This may lead to other lines being sub-optimally sequenced.

Other forms of objective function relating to the individual lines may be formulated which sums the minimum completion times of the individual packing lines.

Table 6.3   Changeover data

SKU	1	2	3	4	5	6	7	8	9
1	0.0	0.5	8.0	8.0	8.0	0.5	0.5	0.5	0.5
2	3.0	0.0	0.5	0.5	0.5	0.0	8.0	0.5	0.5
3	0.5	0.5	0.0	0.5	0.0	8.0	8.0	0.5	0.5
4	0.5	4.0	0.5	0.0	0.0	0.0	0.5	0.5	0.5
5	0.5	0.5	0.5	0.5	0.0	8.0	8.0	8.0	8.0
6	0.5	0.5	0.5	0.5	0.5	0.0	8.0	0.5	0.5
7	0.5	0.5	0.5	0.5	0.5	0.0	0.0	0.5	0.5
8	0.5	0.5	0.5	0.5	0.5	0.0	0.5	0.0	0.5
9	0.5	0.5	0.5	0.5	0.5	0.0	8.0	0.5	0.0

Figure 6.2   Gantt chart of the packing-hall scheduling problem in Example 1

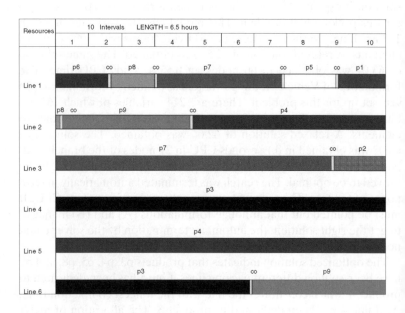

7.2. Example 2

Small-scale packing-hall scheduling problem with nonoverlapping constraints

In some applications, it is required that certain events *do not coincide* at a given time. In this example, it was required that no major changeovers with duration of 8 hours should overlap. To implement this condition other additional constraints are necessary in the basic model. To remove the overlapping of major changeovers (if only one major changeover at any time is allowed), the nonoverlapping constraints need to be imposed. The problem in Example 1 was used to test the algorithm given in formulation (C). Following the reasoning followed in the model, the Vehicle Routing Problem formulation was used to test the algorithm. The resulting model has 399 variables of which 333 are binary variables. The model has 784 constraints with 2950 nonzero entries.

A relaxed solution of 62.66 was obtained followed by an integer solution of 63.23 after examining 23 nodes of the branch-and-bound tree. This solution was obtained in less than 1 second CPU. The search was continued to find the optimal solution for the problem. A best solution of 63.02 was found after examining 29 nodes of the branch-and-bound tree in less than 1 second CPU. The branch-and-bound search was automatically terminated after examining 82 nodes of the tree. The CPU time was less than 2 seconds. The optimal solution indicates that the optimal solution to the problem is similar to the one with the enforced constraints on the nonoverlapping of the changeovers. This proves that the choice of the optimal solution in the basic model satisfies the nonconcurrency conditions imposed on the packing lines.

It is possible to generate the optimal solution using additional constraints in the model. As it was explained earlier, such constraints are indeed redundant, but their inclusion in the problem formulation speeds the convergence to the optimal solution. To obtain the optimal solution, a constraint was added to the basic problem to generate a solution with a completion time of less than 70 hours. After 0.6 seconds CPU, a solution was generated with a completion time of 0.68 seconds. In second run of the model with an additional constraint that the completion time should be less than 68 hours, a solution was generated with a completion time of 64.3 hours. This solution was generated in 1.1 CPU seconds. The third run imposed the completion time of 64.3 as an upper bound and after 1.2 seconds a solution of 63.02 was obtained. No further solution could be generated with a smaller completion time. As it was explained earlier, this solution is optimal and the procedure has found the optimal solution in less than 2 seconds. Although, the user terminated the search, a satisfactory solution (in this case) optimal was found in comparable time.

7.3. Example 3

Application of the technique to the commercial packing-hall model

In a manufacturing environment, the packing hall is connected in operation to other upstream stages. In this case study, the factory has 12 packing lines of different capabilities and packing speeds. Some lines are similar in operation and are run as members of groups of lines related in operation (in practice, this means that packing lines receive similar product variants). The demands, variants, and pack sizes for all SKUs are given in Table 6.4 and the packing rates in Tables 6.5 and 6.6. In this case study, there are 48 items to be packed.

There is a severe sequence dependency betweem respective SKUs on packing lines. The changeover times range between 12 and 0.5 hours.

Table 6.4   Demands, pack sizes, and variants for first 24 SKUs in the case study

Item	Variant	Pack Size	Demand	Item	Variant	Pack Size	Demand
1	1	E	19.00	25	9	H	41.00
2	2	B	12.00	26	9	K	26.00
3	2	D	52.00	27	9	L	32.00
4	2	E	27.00	28	9	F	56.00
5	3	D	47.00	29	11	I	53.00
6	4	C	20.00	30	12	I	18.00
7	4	E	41.00	31	12	M	68.00
8	4	F	15.00	32	12	M	68.00
9	6	H	21.00	33	13	M	20.00
10	7	A	20.00	34	14	I	96.00
11	7	B	49.00	35	14	H	516.00
12	7	C	22.00	36	14	H	31.00
13	7	G	26.00	37	14	K	283.00
14	7	D	181.00	38	14	K	39.00
15	7	D	35.00	39	14	N	46.00
16	7	D	23.00	40	15	I	19.00
17	7	F	58.00	41	15	J	30.00
18	7	F	3.13	42	15	H	202.00
19	7	F	3.13	43	15	K	167.00
20	7	F	2.00	44	15	K	41.00
21	8	B	29.00	45	16	I	119.00
22	9	I	45.00	46	16	H	641.00
23	9	I	24.00	47	16	K	326.00
24	9	J	35.00	48	16	N	57.00

Table 6.5   Packing rates for first 24 SKUs in the case study

	L1	L2	L3	L6	L7	L8	L9	L10	L4	L11	L12	L13
1						3.15	3.15	3.15				
2				5.94								
3						2.52	2.52	2.52				
4						3.15	3.15	3.15				
5						2.52	2.52	2.52				
6		5.28										
7						3.15	3.15	3.15				
8										8.4	16.8	16.8
9	14.0											
10				1.98								
11				3.17								
12			2.89									
13			4.16									
14						1.26	1.26	1.26				
15						1.51	1.51	1.51				
16						1.51	1.51	1.51				
17										5.0	10.0	10.0
18										5.0	10.0	10.0
19										5.0	10.0	10.0
20										5.0	10.0	10.0
21				3.43								
22		2.97										
23		2.97										
24				4.62								

The minimum run length is not a constraint as far as the operation of this factory is concerned. The details of changeovers on the packing lines are as follows. The sequence-dependent changeovers on line 3 are asymmetric and amount to 4 hours between pack sizes C and G and 12 hours between C and H, as well as between G and H. There are also major changeovers of 12 hours between pack sizes K and L on lines 7 and between H and K on line 1. The sequence-dependent changeovers on line 6 are shown in Table 6.7.

Lines 8, 9 and 10 changeovers are as follows. All changeovers including pack size changeovers between D and E are usually of 1 hour duration, except the label changeovers, which are of 0.5 hour duration. All changeovers on lines 11, 12, and 13 were assumed to be zero. For all packing lines in general, all variant changeovers are 1 hour duration and all label changeovers on packing lines are of 0.5 hour duration.

Table 6.6   Packing rates for last 24 SKUs in the case study

	L1	L2	L3	L6	L7	L8	L9	L10	L4	L11	L12	L13
25	5.61											
26					4.95							
27					4.95							
28										4.25	8.5	8.5
29		4.1										
30		6.9										
31				7.0								
32				5.6								
33				5.6								
34		6.9										
35	14.0		13.2									
36	11.9		11.2									
37	12.1				7.92							
38	12.1				7.92							
39												
40		6.93										
41				8.09								
42	14.0		13.2									
43	12.1				7.92							
44	12.1				7.92							
45		6.93										
46	14.0		13.2									
47	12.1				7.92							
48									5.32			

Table 6.7   Line *L*6, sequence-dependent major changeovers

Line 6	A	B	M	J
A	0	12	8	12
B	12	0	8	8
M	8	8	0	8
J	12	8	8	0

The problem is defined as follows. Schedule the operation of lines as far as the 48 SKUs are concerned to achieve a minimum packing span within the available time of 120 hours. The different formulations of (A), (B), and (C) as outlined earlier in this chapter were evaluated in this commercial case study. Initially, formulation (A) was set up. The problem representation contains 540 active decision variables of which

369 are binary variables. There are 1382 active constraints in this formulation. There are 3216 nonzero elements in the constraints. A relaxed solution of 83 was obtained in 1 second CPU time. A solution of 116.5 was obtained in 2301 nodes of the branch-and-bound tree. The solution was obtained in 60 seconds CPU. No other solution of lower completion time for the packing hall could be generated. Simple analysis shows that this solution is sub-optimal. Formulation (B) with better lower-bound properties was set up for this case study. This formulation has 824 variables with 653 binary variables. There are 1737 constraints with 6146 nonzero elements. The relaxed objective value of 83 was obtained in 413 iteration of simplex. As before, a solution of 116.2 was obtained in 1861 nodes of the branch-and-bound tree. This result was obtained in 48 seconds CPU. Although the lower-bound and the first-integer solution did not improve, the integer solution was generated in faster CPU time. No other solution could be generated within the limit of 10 000 nodes reached. Finally the Vehicle Routing Problem formulation (C) was set up for this commercial packing-hall problem. This formulation has more variables than formulation (B); however, there are less constraints. There are 994 variables of which 823 are binary. There are also 1079 constraints with 5430 nonzero elements. A solution of 105.2 was obtained in 3 seconds CPU in 161 nodes of branch-and-bound tree. After generating solutions of 104.3, and 101.7 in 48 and 104 seconds CPU a solution of 101.1 was obtained in 373 seconds CPU. The number of nodes examined was 19 664. No other solutions were generated and the search was terminated automatically in less than 10 minutes. Examination of the results reveals that the optimal solution corresponding to the problem has been obtained. The Gantt chart of the solution is given in Figure 6.3. It must be pointed out that the uniform time representation of the events would result in a very large number of variables (binary and continuous). Other continuous-time formulations would result in twice as many binary variables as well as a large number of binary continuous variables.

To prove the optimality of the solution, the following explanation is necessary. Examining the structure of the problem reveals that the scheduling problem consists of two groups of lines, which are unrelated in operation. Group 1 consists of Line $L6$ only while other packing lines belong to group 2. Clearly, the scheduling of each group can be carried out independent of each other if the packing hall is scheduled independent of the factory as a whole. In reality, the packing lines can not be grouped independently. The solution completion time for group

1 is 101.2. This solution corresponds to the optimal solution. This can proved as follows. The sequence generated for this line is given below

$$p10 \rightarrow p33 \rightarrow p32 \rightarrow p31 \rightarrow p24 \rightarrow p41 \rightarrow p21 \rightarrow p11 \rightarrow p2 \quad (6.47)$$

Apart from the minor changeovers between these products, there are major changeover requirements categorized by A, M, J, and B. The above product sequence as outlined by the sequence in (6.46) corresponds to size changeovers

$$A \rightarrow M \rightarrow J \rightarrow B \quad (6.48)$$

Examination of Table 6.7 reveals that there are several possible optimal solutions. This table also reveals that an optimal sequence should contain a minimum changeover between A and M (from A to M), a minimum changeover between B to M or J (from B to M or J), and from M to any other size and also from J to B or M. Hence the above sequence is an optimal sequence.

Consider the group 2 lines in the solution. Examination of the results reveals that the solution is optimal with regard to the minimum overall completion time in this group of lines in the packing hall. The solution for this group of lines is bounded by line 3.

The sequence for this line is generated as below

$$p6 \rightarrow p12 \rightarrow p13 \rightarrow p35 \rightarrow p36 \rightarrow p42 \rightarrow p46 \quad (6.49)$$

The above product sequence corresponds to the *size* changeovers given below

$$C \rightarrow G \rightarrow H \quad (6.50)$$

This sequence is optimal.

To enforce the nonconcurrency of the changeovers on the results of this case study, consider the results given in Gantt chart of Figure 6.3. The lines $L1$ and $L6$ are changeovers between hours 65 and 69. In order to remove the overlapping of the changeovers, additional constraints were added to the model. To remove the overlapping major changeovers, the Vehicle Routing Problem formulation and its extension as given by relationships (6.34), (6.35), (6.36), (6.37) were added to the model. The new model has 6104 rows, 3015 decision variables of which 2789 are binary variables. After 162 nodes of the branch-and-bound tree with a CPU time of approximately 1600 CPU seconds, a solution with the objective value similar to the base model was generated. The

Figure 6.3   Gantt chart of the commercial packing-hall scheduling problem
in Example 3

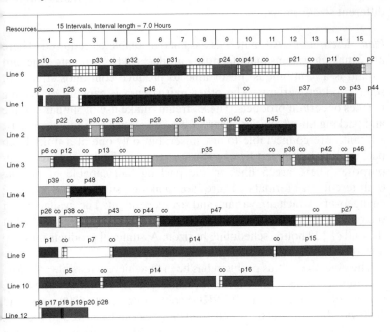

resulting solution is essentially the same, except that the sequence of
the SKUs on line 1 has been changed to the following. The sequence
starts from *p*37 to *p*44 followed by a changeover of 12 hours and a
sequence of SKUs *p*9 to *p*46. The search was terminated in less than
2000 seconds CPU.

7.4. Example 4

Multi-stage consumer goods' liquids manufacturing problem

The problem objective is to schedule the manufacturing of a portfolio
of products in a factory using all the resources available on site.
The factory works for 120 hours a week (5 days). The schematic-
manufacturing route is given below,

   Making $\Rightarrow$ Intermediate Storage $\Rightarrow$ Packing $\Rightarrow$ Warehousing

The objective may be assumed to be the minimization of the factory completion time. There are 28 SKUs in 10 different variants and in 6 different size/formats.

Plant description   All materials assumed available as required. There are 2 making systems manufacturing products in two streams. One unit stream is a 'continuous' mixer able to produce continuos stream of some products/variants. The other unit stream is capable of making batches of certain products. These streams make use of similar storage and packing lines/facilities. There are 4 tanks, each of a finite capacity. All tanks are connectable to any mixer, but only one at a time. Each packing line is only fed by one storage resource. The tanks are multi-purpose. There are 5 lines in the packing hall capable of making 6 different size/formats. There are making, storage and packing changeovers, which are variant-and size-dependent. The availability of the packing hall is 120 hours. All product demands must be satisfied at the end of 120 hours' scheduling horizon. Assume all products move to a warehouse from the packing hall.

The objective of this problem has been to achieve a feasible solution. Different versions of the model have different CPU solution times. The version of this model with 18 SKUs containing nearly 5000 rows, and 1870 variables of which around 1000 binaries can be solved to the first feasible solution in 23000 nodes with a CPU of 10 minutes on a Pentium Pro. 200 MHz platform.

VIII. CONCLUSIONS

Most work on the detailed scheduling of packing halls in the presence of complex constraints resorts to discretizing time. The resulting models have a large number of binary variables to an extent that rules out solving them by standard packages. Moreover, the solutions achieved through specialized procedure tend to exhibit sub-optimality due to the approximation inherent in the discretization process itself. In contrast, the model presented here is essentially continuous-time, in that it allows all events, whether packing, changeover or idle, to start and end at any time. The model simplicity makes it possible to handle common resource constraints due to material supply (raw material or semi-finished products), manpower or any concurrency of events limitation.

The modeling process is thereby complicated, but the resulting model tends to be compact with a relatively small number of binary variables and constraints. This makes it possible to employ standard solvers, such as CPLEX or XPRESS-MP, the state-of-the-art mathematical programming package.

The computational experiments suggest that the approach presented in this work, has the potential to outperform other methods, because of the simplicity of the logical decisions involved and the efficiency of the formulation. This approach allows the solution of real large-scale design/scheduling problems in the process industry in an automatic manner. The Gantt charts of the solutions can also be generated automatically.

Note

The author wishes to thank Professors L. Wolsey, Y. Pochet of CORE (Center for Operations Research and Econometrics), Université Catholique de Louvain, Belgium, and K.H. Hindi of Brunel University, UK for their valued suggestions.

References

Driscoll, W. and H. Emmons (1977) 'Scheduling Production on One Machine with Changeover Costs', *AIIE Transactions*, 9, 388–95.

Geoffrion, A.M. and G.M. Graves (1976) 'Scheduling Parallel Production Lines with Changeover Considerations', *Operations Research*, 24, 595–610.

Glassey, C.R. (1968) 'Minimum Changeover Scheduling on One Machine', *Operations Research*, 16, 343–53.

Gooding, W.B., J.F. Pekny and P.S. McCroskey (1994) 'Enumerative Approaches to Parallel Fellowship Scheduling via Problem Transformations', *Computers and Chemical Engineering*, 18, 909–27.

Kondili, E., C.C. Pantelides and R.W.H. Sargent (1993) 'A General Algorithm for Short term Scheduling of Batch Operations – I.MILP Formulation', *Computer and Chemical Engineering*, 17, 211–28.

Love, R.R., Jr and R.R. Vegumanti (1978) 'The Single Plant Old Allocation Problem with Capacity and Changeover Restrictions', *Operations Research*, 26, 156–65.

Moore, J. (1975) 'An Algorithm for a Single-machine Scheduling Problem with Sequence Dependent Setup Times and Scheduling Windows', *AIIE Transactions*, 7, 35–41.

Musier, R.F.H. and L.B. Evans (1989) 'An Approximate Method for Production Scheduling of Industrial Batch Processes with Parallel Units,' *Computers and Chemical Engineering*, 13, 229–38.

Parker, R.G., R.H. Deane and R.A. Holmes (1977) 'On the Use of a Vehicle Routing Algorithm for the Parallel Processor Problem with Sequence Dependent Changeover Costs', *AIIE Transactions*, 9, 155–60.

Pinto, J.M. and I.E. Grossmann (1995) 'A Continuous Time Mixed Integer

Linear Programming Model for Short Term Scheduling of Multistage Batch Plant', *Industrial Engineering Chemistry Research*, 34, 3037–51.

Pinto, J.M. and I.E. Grossmann (1996) 'A Continuous Time MILP Model for Short Term Scheduling of Batch Plants with Pre-ordering Constraints', *Computers and Chemical Engineering*, 29, S1197–S1202.

Prabhakhar, T. (1974) 'A Production Scheduling Problem with Sequencing Considerations', *Management Science*, 21, 34–43.

Reklaitis, G.V. (1991) 'Perspective on Scheduling and Planning of Process Operation', *Proceedings of the Fourth International Symposium on Process Systems Engineering*, Montreal.

Reklaitis, G.V. and L. Mockus (1995) 'Mathematical Programming Formulation for Scheduling of Batch Operations on Non-uniform Time Discretisation', *Acta Quimica Slovenica*, 42, 81–6.

Robert, L., Jr. (1976) 'Sequencing with Setup Costs by Zero–one Mixed Integer Linear Programming', *AIIE Transaction*, 8, 369–71.

Sahinidis, N.V. and I.E. Grossmann (1991) 'MINLP Model for Cyclic Multipurpose Scheduling on Continuos Parallel Lines', *Computers and Chemical Engineering*, 15, 85–103.

Schilling, G. and C.C.A. Pantelides (1996) 'Simple Continuos Time Process Scheduling Formulation and a Novel Solution Algorithm', *Computers and Chemical Engineering*, 20,S–B, S1221–S1226.

Voudoris, T.V. and I.E. Grossmann (1993) 'Optimal Synthesis of Multiproduct Batch Plants with Cyclic Scheduling and Inventory Consideration', *Industrial Engineering Chemistry Research*, 32, 1962–80.

Zentner, M.G. and G.V. Reklaitis (1992) 'An Interval-based Mathematical Model for the Scheduling of Resource-Constrained Batch Chemical Processes', *Proceedings of NATO. ASI on Batch Processing Systems Engineering*, Antalya.

Zhang, X. and R.W.H. Sargent (1994) 'The Optimal Operations of Mixed Production Facilities – A General Formulation and Some Solution Approaches for the Solution', *Proceedings of the 5th International Symposium on Process Systems Engineering*, Kyongju, 171–7.

7 Solving Airline-fleet Scheduling Problems with Mixed-integer Programming

Uwe H. Suhl and Leena M. Suhl

I. INTRODUCTION

This chapter discusses the solution of real-life airline fleet scheduling problems of a large European airline, and aims to minimize the number of aircraft needed to serve a given set of flights. There is certain freedom to schedule the flights expressed as a time window per each flight within which the flight has to depart. This is the strategy of some European carriers, which first fix the number of flights per connection together with the time window and the fleet for each flight according to the expected number of passengers. In a second step they schedule the flights within their given departure time windows. This problem of scheduling the flights and simultaneously generating aircraft rotations will be called the *fleet scheduling* problem. (For more details on the planning process see Suhl, 1995.)

Input data to the fleet scheduling problem are expressed as a tentative set of *products* describing all flights to be served within one week.

The product of a commercial airline – a flight – is a transportation service defined by

- a time period (for example summer or winter)
- days of a week (for example, daily except Sundays)
- departure and arrival times
- departure and arrival airports
- capacity (expressed as an aircraft pattern or fleet).

In case of tentative products the departure and arrival times are not fixed but given as time windows, typically having a duration of a few hours. Generally, a production scheduling process is needed at an airline to generate a flight schedule and assign resources to it. It

135

Figure 7.1   The rotation generation procedure for an Airbus 312

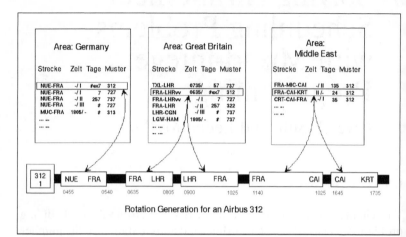

Rotation Generation for an Airbus 312

typically involves the following four basic phases: *product planning* (the connections to be served by an airline), *production planning and scheduling* (fleet assignment, aircraft routing and flight scheduling), *resource allocation* (assigning necessary resources, like crews, physical aircraft, and airport gates), and *operation time rescheduling* (rescheduling flights or crews because of nonpredictable changes). In this paper, we concentrate on the production planning and scheduling phase in such a way that aircraft routing and flight scheduling are performed simultaneously.

Here we consider a long-term planning problem of scheduling tentative flights. The solution generates *weekly rotations* for logical aircraft. A later phase decides which physical aircraft will take over a certain logical rotation. Each flight has to depart within a given time window and between two flights a minimum standing time (ground time) at the airport has to be guaranteed. A time window may extend over a few hours, it may have the length zero (the departure time being fixed), or it can stretch over one day or several days. The weekly scheduling problem may contain flights to be carried out only on certain weekdays.

In the industrial routing and scheduling process, product managers, each of which is possible for a certain geographic area (Figure 7.1), produce a draft schedule. The product managers estimate the number of flights for each connection and day of week and suggest the aircraft pattern and time window. The ultimate goal is to generate a flight schedule which maximizes profit. However, since the profit of a certain flight is difficult to estimate exactly in long-term planning, the goal is approximated by the minimization of the number of aircraft needed to carry out all given flights. In an early planning stage, the generated rotations can be used to determine the needed fleet size and mix. In our model, maintenance activities are implicitly modeled as 'dummy' flights.

The problem can be modeled as a linear minimum cost network optimization problem where each flight corresponds to a node. An arc is generated between nodes being potential predecessor–successor– pairs carried out by the same aircraft. Additionally, nodes labeled SOURCE (number 0) and SINK (number $n + 1$) are generated so that all new aircraft taken into the system emanate from SOURCE and end in SINK (Figure 7.2).

II. HEURISTIC ALGORITHMS

We first approached the airline fleet scheduling problem with heuristic methods to generate a reasonably good solution fast. This solution is then used to guide the solution process within the optimization process. The heuristics used here are a modification of well known techniques for the case where all departure times have been fixed.

In the routing problem with *fixed departure times*, there are usually many possible successors and predecessors of a flight (see Figure 7.2). Three basic criteria can be used to select a successor flight:

- **first-in-first-out** (FIFO): the aircraft which arrived first, will leave first
- **last-in-first-out** (LIFO): the aircraft which arrived last, will leave first
- **best-first** (BF): choose that flight from available (not yet planned) flights which can be flown by an aircraft with the shortest standing time.

Although all three methods give the minimum fleet size for the problem with fixed departure times, they produce different rotations. The first two heuristics proceed the flights in increasing order according to departure times. The FIFO strategy tends to use aircraft more evenly: since the aircraft standing longest will leave first, the fluctuation in

Figure 7.2   Aircraft routing as a network flow problem (schematically)

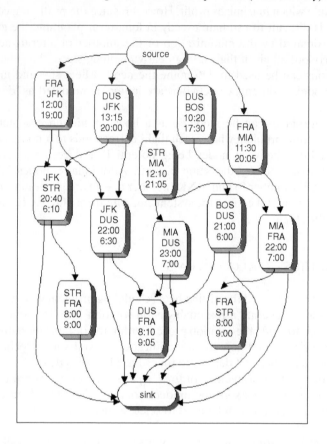

standing times remains relatively small. This is desirable in the case where all flights for a pattern are given. However, at early planning stages we often want to add new flights or change the pattern for some given flights. In such cases the LIFO rule is more desirable, because then the standing times are unevenly distributed. It may be possible to insert a new flight in a large „hole" within a rotation. The best-first algorithm is at best performed in an *aircraft oriented* way: We first construct a complete rotation for one aircraft, then for the second one, using the flights that are left, and so on. It is obvious that the first rotation will be a good one having small standing times between the flights, and that the rotations deteriorate in the order they are generated.

In order to generate reasonably fast good solutions to the routing problem *with time windows*, we modified the three heuristics in the following way (for details, see Suhl, 1995). For the LIFO and FIFO strategy we first sort the flights in increasing order according to first possible departure times. We assign the flights to aircraft in this order, so that all flights which started before a given time window are known. These algorithms work in the following way:

Initially: We have no aircraft in use; there is a pool of identical aircraft. We process the flight list sequentially. For each flight either (**a**) or (**b**) is true:

(**a**) **There is an aircraft available** at the requested airport within the time window (either there is at least one in the beginning or one will become available before the end of the time window). If there is only one possible aircraft, we assign it; otherwise we choose one according to the given (LIFO or FIFO) strategy.

(**b**) **There is no aircraft available**. Normally we take a new aircraft from a pool. If there are several tentative flights (one of which is to be selected) we take a new aircraft only if this is the last tentative flight within a day. Tentative flights correspond to a completely free departure time within one day that has been divided into several alternative moderate-size time windows. The heuristics work properly only if the differences between time windows are not too large.

The LIFO and FIFO strategies process the sorted flights sequentially. They differ in the case where several aircraft are available for one flight. The FIFO strategy chooses the aircraft which arrived (or will arrive within the time window) first. The LIFO strategy chooses the aircraft which arrived as late as possible, thus generating larger gaps where new flights may be inserted.

The third method, the aircraft oriented best-first principle, works like the best-first method without time windows. We start with one aircraft and generate a rotation having as little slack as possible – taking the time windows into account. Simultaneously we mark the flights, which have been assigned. Among the remaining flights we generate a second rotation similarly, and so on. This method has proven very useful for practical problems, since it reveals flights, which fit poorly together with other flights. The flight scheduler may then want to change the time window or itinerary of such flights.

III. THE OPTIMIZATION MODEL

A mathematical formulation of the weekly problem with n flights, or tasks, can be stated:

$$\min \sum_{i=0}^{n+1} \sum_{j=0}^{n+1} c_{ij} x_{ij} \tag{7.1}$$

subject to:

$$\sum_{k=0}^{n} x_{kj} = \sum_{k=1}^{n+1} x_{jk} = 1 \quad j \in \{1, \ldots, n\} \tag{7.2}$$

$$x_{ij} (T_j - T_i - t_i) \geq 0 \quad i \in \{1, \ldots, n\}, \; j \in \{1, \ldots, n\} \tag{7.3}$$

$$a_i \leq T_i \leq b_i \quad i \in \{1, \ldots, n\} \tag{7.4}$$

$$x_{ij} \in \{0, 1\} \quad i \in \{0, \ldots, n\}, \; j \in \{1, \ldots, n+1\} \tag{7.5}$$

Here x_{ij} is a binary decision variable: $x_{ij} = 1$, if flight j follows directly after flight i with the same aircraft; otherwise $x_{ij} = 0$. T_i is a continuous decision variable denoting the departure time of flight i within a time window. The constant t_i represents the duration of flight i and contains the minimum ground time before a successor flight can start. The cost coefficients are denoted by c_{ij}: if our only goal is to minimize the fleet size, we set the cost of flights emanating from SOURCE equal to 1, and all other costs equal to zero. This was done for the application which will be reported below.

Constraints (7.3) relate the tasks with each other: the next task with a resource (in our case, airplane) can be started only after the previous task is finished. The constraints (7.4) define time windows for all tasks in the network. Note that this formulation has a quadratic term. A linear reformulation is possible with the „big-M" method:

$$T_j - T_i - t_i \geq M_{ij}(x_{ij} - 1) \quad i \in \{1, \ldots, n\}, j \in \{1, \ldots, n\} \tag{7.6}$$

Each coefficient M_{ij} is individually chosen for every pair (i, j) so that its value is as small as possible depending on a_i, b_i, a_j, b_j, and t_i. This helps us to tighten the LP-relaxation involved. In addition as will be explained in sub-section 4.2, a coefficient reduction as part of integer preprocessing will be used to reduce the size of those coefficients.

The problem (7.1) . . . (7.5) is sometimes referred to as the *multiple traveling salesman problem with time windows* (mTSPTW). If the only objective to be minimized is the number of aircraft, this problem can be referred to as the *minimum fleet size multiple traveling salesman problem with time windows.*

The minimum fleet size problem (7.1) . . . (7.5) arises in several other contexts, such as school bus routing, urban transit, passenger trains, and assignment to *m* machines of *n* tasks which must be carried out within a given time interval.

In general, the literature on vehicle routing and scheduling with time windows has been rapidly growing over the last 15 years. The traveling salesman problem with time windows has been studied in Baker *et al.* (1995) and Christofides *et al.* (1979). For the vehicle scheduling problem with time windows, heuristic methods optimizing fleet size and routing costs have been developed in Bodin *et al.* (1983) and Desrosiers *et al.* (1984). With the time window discretized, optimal algorithms for the fleet size have been designed using integer programming formulations in Levin (1971) and Gertsbach and Gurevich (1977). An exact algorithm proposed by Desrosiers *et al.* (1984) uses a set covering formulation solved by a column generation scheme where the subproblem is a shortest-path problem with time window constraints. Desrosiers *et al.* (1991) propose a column generation method which has proven useful and efficient for many practical problems, especially when time windows are relatively small.

In this chapter, we discuss the solution of the *m*TSPTW problem by standard software for mixed integer programming. We show that practical problems can be solved by this approach. It has the advantage of being general – in other words, even if some constraints are added which change the problem structure, the approach can still be used, in contrast to dedicated heuristic methods.

IV. SOLVING THE MIXED-INTEGER MODEL

Using matrix notation we may state a mixed-integer optimization problem in the following form:

$$\min \quad c'x$$
$$bl \leq Ax \leq bu \tag{P}$$

$l \leq x \leq u$

x_j integer for $j \in J_l$

where l, u, c, x are n-vectors, bl, bu are m-vectors, and A is a spare m× n coefficient matrix. All coefficients are assumed rational except that the vectors l, u, bl and bu may contain elements that are plus or minus infinity. $I = \{1, \ldots, m\}$ and $J = \{1, \ldots, n\}$ are the index sets of constraints and variables. $J_I \subseteq J$ is the index set of integer variables, $J_{01} \subseteq J_I$ is the index set of zero–one variables. The jth component of a vector v is denoted by v_j; a_{ij} denotes the matrix element in row i and column j of A, and a_j the column j of A.

The LP-relaxation (LP) of (P) is defined as the optimization problem where the integrality constraints for the integer variables are dropped. $F(P)$ denotes the set of feasible points of (P), $F(LP)$ the set of feasible points of the LP-relaxation (LP).

All successful solution algorithms for solving (P) are based on branch-and-bound algorithms where at each node of the tree the LP-relaxation (LP) of the corresponding sub-problem is solved. Fathoming of a sub-problem occurs only if (LP) is infeasible or integer or if the objective function value $z(LP)$ of (LP) is worse than the best integer solution found so far. As a consequence the representation of (P) by the underlying LP-relaxation is critical.

Significant algorithmic progress for solving integer models (P) during the last decade is based on the notion of *tight linear programming relaxation*: an important contribution comes from cutting planes (facets or faces of high dimensions) derived from (P) which are added resulting in an integer equivalent problem (P') with tighter (LP') than (LP). The concept of tight LP-relaxation originated from E.L. Johnson in a joint research project between IBM Research and General Motors to solve difficult pure 0–1 problems (Crowder *et al.*, 1983; Johnson *et al.*, 1985). The research results have been implemented in the IBM solver Optimization Subroutine Library (1989). The tight LP-relaxation is now one of the important research topics in integer programming (Johnson, 1989; Wolsey, 1989; Hoffman and Padberg, 1991; Dietrich *et al.*, 1993).

Experience shows that the successful application of integer programming in practice requires:

- **A good model formulation** with emphasis on a tight LP-relaxation.
- **A state-of-the-art MIP**-optimizer with

- **Algorithms based on branch-and-bound** where specific sub-problems (nodes) of the search tree are processed to tighten the LP-relaxation. These nodes are named *supernodes*. The initial problem (*P*) is always a supernode.
- **Fast and stable simplex-based LP-optimizer** with primal and dual algorithms to re-optimize the LP-relaxation. The simplex algorithm is the method of choice within branch-and-bound to take advantage of basic solutions (tight LP-relaxation) and starting bases.
- **Efficient system design and implementation** where advantage is taken from the large main memories of workstations and high-end PCs.
- **Model-dependent branching and node selection strategies** which capture inside knowledge of the underlying practical application. This is very important for difficult models in particular with time dependent sequencing decisions.

All of these aspects are of crucial importance in solving the fleet scheduling problem discussed in the previous sections.

4.1. The optimization solver

MOPS® (*M*athematical *Op*timization *S*ystem) is a mathematical programming software system for solving large-scale LP and mixed integer optimization problems (Suhl, 1994; Suhl and Szymanski, 1994). The system is used in industrial applications on system platforms ranging from PCs to mainframes (MVS). The core of the system is a high-speed simplex-based primal and dual LP-optimizer. The integer models are solved with a branch-and-bound optimizer with general node selection strategies and efficient node-buffering on disk. All MOPS routines are part of an object code library. Under Microsoft Windows (Windows '95, Windows NT) the MOPS Dynamic Link Library (DLL) allows access of optimization functions from programs written in Visual Basic, C++, and Delphi. The MOPS-library has been successfully used in branch-and-cut algorithms to solve special purpose optimization problems such as generalized Steiner graph problems (Suhl and Hilbert, 1996), routing problems with column generation, and set partitioning problems.

To tighten the LP-relaxation (supernode processing) MOPS uses well known techniques, described in the literature. Among them are:

- feasibility testing and fixing of variables (Brearley, Mitra and Williams, 1975)
- bound reduction, constraint inactivation, duality tests (Brearley, Mitra and Williams, 1975)
- coefficient reduction (Johnson *et al.*, 1985; Savelsbergh, 1994)
- logical testing and probing (Guignard and Spielberg, 1981)
- recognition of cliques and implications (Dietrich *et al.*, 1993)
- generation of clique and implication cuts (Johnson, 1989; Suhl and Szymanski, 1994)
- special treatment of fixed-charge variables (Suhl, 1985)
- generation of cover constraints (Crowder *et al.*, 1983)
- Euclidean reduction (Hoffman and Padberg, 1991)

For a selected sub-problem of (P) (a supernode) where some variables are fixed or have changed bounds and some constraints are marked as redundant, these techniques are applied in a specific sequence to tighten the LP-relaxation as shown in Figure 7.3. The m-vectors L and U define for each constraint a lower and upper bound for the corresponding value. L and U may have to be updated when additional variables are fixed, coefficients or bounds of variables are changed (Figure 7.3). (Suhl and Szymanski, 1994). As default MOPS performs supernode processing only for the initial problem (P) then an internal heuristic finds an integer solution (see also Figure 7.4). This approach was taken to solve the fleet scheduling problems.

4.2. Probling, implications and extended coefficient reduction

One particular preprocessing technique is the extended coefficient reduction, because of its critical importance for solving fleet scheduling problems; it will be discussed in greater detail in conjunction with probing and implications.

Let $S \subseteq J$ the index set of fixed variables, i.e. if $j \in S$ then $x_j = v_j$, where $l_j \leq v_j \leq u_j$. $F(S) = J - S$ is the index set of non-fixed variables. We start initially with $S = \emptyset$. A globally fixed variable is recorded in S. The m-vector $a(S) = \Sigma_{j \in S} v_j \, a_j$ reflects the impact of fixed columns to the right-hand side vectors. Let $P_i = \{j \mid j \in J, a_{ij} > 0\}$ and $N_i = \{j \mid j \in J, a_{ij} < 0\}$. Lower and upper row sums $L(S)_i, U(S)_i$ for row i can be computed as follows:

$$L(S)_i = a(S)_i + \sum\nolimits_{j \in P_i \cap F(S)} a_{ij} l_j + \sum\nolimits_{j \in N_i \cap F(S)} a_{ij} u_j$$

Figure 7.3   Supernode processing in MOPS

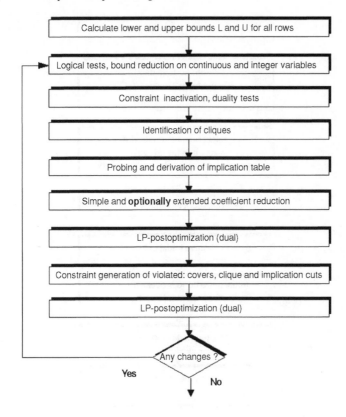

$$U(S)_i = a(S)_i + \sum\nolimits_{j \in P_i \cap F(S)} a_{ij} u_j + \sum\nolimits_{j \in N_i \cap F(S)} a_{ij} l_j$$

The lower and upper sums play a key role for detection of infeasibility, redundancy, fixing of variables, bound and coefficient reduction.

Infeasible and redundant rows

If $L(S)_i > bu_i$ or $U(S)_i < bl_i$ then the sub-problem defined by S is infeasible. If $L(S)_i > bl_i$ and $U(S)_i < bu_i$ then the ith constraint is redundant.

Figure 7.4 The overall system to solve fleet scheduling problem

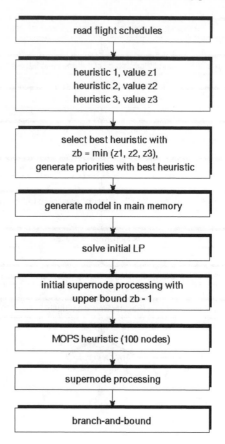

Degree 1 tests (only for 0–1 variables)

If $j \in J_{01} \cap F(S)$ and $L(S)_i + |a_{ij}| > bu_i$ then variable j can be fixed. If $j \in P_i \cap J_{01}$ then variable j can be fixed to 0 and if $j \in N_i \cap J_{01}$ variable j can be fixed to 1. If $j \in J_{01} \cap F(S)$ and $U(S)_i - |a_{ij}| < bl_i$ then variable j can be fixed. If $j \in P_i \cap J_{01}$ then variable j can be fixed to 1 and if $j \in N_i \cap J_{01}$ variable j can be fixed to 0.

Probing is a special form of a degree 2 test (Guignard and Spielberg, 1981). The idea is to set a nonfixed 0–1-variable tentatively to 0 or 1. Degree 1 tests are then applied to all nonfixed variables (not only to

0–1 variables). If any additional variables can be fixed the process is repeated. MOPS stores initial probing implications as a table of signed indices together with cliques (cliques are symmetric implications) (Suhl and Szymanski, 1994). Probing is very important for the following reasons:

- probing might fathom the current sub-problem – i.e. prove that it is infeasible
- the implications can also be used to fix variables if there are contradictions
- implications are algebraically equivalent to cuts (see below); if they are violated by the current LP-solution they can be used as cutting planes
- implications can also be used to tighten bounds of continuous variables (*conditional bounds*)
- implications from probing can also be used for an extended coefficient reduction.

Logical implications derived by probing imply inequalities which can be used as cutting planes. Let $l \leq x \leq u, y \in \{0, 1\}$ where x is continuous or integer. It is easy to see that the following implications imply the corresponding inequalities called implication cuts:

$$(y = 1 \Rightarrow x = a) \Rightarrow \begin{cases} x \geq l + (a - l)\, y \\ x \leq u - (u - a)\, y \end{cases}$$

and

$$(y = 0 \Rightarrow x = a) \Rightarrow \begin{cases} x \geq a - (a - l)\, y \\ x \leq a + (u - a)\, y \end{cases}$$

Probing in conjunction with bound reductions can be used to derive conditional bounds also for continuous variables (Savelsbergh, 1994). MOPS uses such conditional bounds also during the branch-and-bound process. The conditional bounds are stored in the implication table. The original bounds are restored if necessary from the saved original bounds.

Extended coefficient reduction

The basic idea of coefficient reduction is to set a 0–1-variable tentatively to 0 or 1 and to test then if a constraint becomes strictly inactive.

If this is the case, the magnitude of the coefficient of the variable tested and possibly the right hand side can be decreased so that the constraint becomes just binding without changing the set of feasible 0–1-variables (Johnson *et al.* 1985). Less well known is the possibility of introducing new nonzeros not present in the original coefficient matrix. Various papers deal with improvements of the original coefficient reduction by taking other constraints, typically special ordered sets, cliques, or variable upper bounds into account (Hoffman and Padberg, 1991; Dietrich and Escudero, 1990). The disadvantage of such an approach is that special structures in the matrix have to be identified and used to calculate tighter L-, resp. U-vectors. Changes in the matrix and fixed variables complicate the update of L, resp. U. MOPS uses an extended coefficient reduction technique which is based only on probing and uses given cliques and implications derived from all constraints in the model. Only the values of $L(S)$ and $U(S)$ are updated and used for the coefficient reduction. This approach is a byproduct of probing. Special structures such as cliques are indirectly exploited by using the information in the clique and implication table to update $L(S)$ and $U(S)$. This type of coefficient reduction may eliminate nonzero coefficients, reduce the magnitude of coefficients or introduce new nonzero coefficients not present in the original coefficient matrix. For $j \in J_{01} \cap F(S)$ let S_j^1 resp. S_j^0 be the set of variables in S including j and the additional variables which can be fixed by setting x_j to 0 or 1 – i.e.

$$S_j^1 = \{j\} \cup S \cup \{k \,|\, k \in F(S) - \{j\}, x_k = v_k \text{ if } x_j = 1\}$$

$$S_j^0 = \{j\} \cup S \cup \{k \,|\, k \in F(S) - \{j\}, x_k = v_k \text{ if } x_j = 0\}$$

The sets S_j^0, S_j^1 are a result of probing on variable j. We consider here the case of a $\leq$ constraint. If $a_{ij} \geq 0$ and $U(S_j^0)_i < bu_i$ let $\delta_{ij} = bu_i - U(S_j^0)_i$. The coefficient a_{ij} can be reduced to $a_{ij}' = \min\{a_{ij} - \delta_{ij}, 0\}$ and the right-hand side has to be changed to $bu_i' = bu_i - \delta_{ij}$. If $a_{ij} < 0$ and $U(S_j^1)_i < bu_i$ the coefficient a_{ij} can be reduced to $a_{ij}' = \min\{a_{ij} + \delta_{ij}, 0\}$. If $a_{ij} = 0$ and $U(S_j^1)_i < bu_i$ then a_{ij} can be changed to $a_{ij}' = bu_i - U(S_j^1)_i$. Similar rules are valid for a $\geq$ constraint.

4.3. Example

The following IP-model is a modified example from Savelsberg (1994). The constraints R7 and R8 were added. Note that the x-variables are 0–1 and the y-variables continuous. The example is used to demonstrate the effect of probing and the extended coefficient reduction as

Table 7.1 Solving the model

Example	x_1	x_2	x_3	y_1	y_2	y_3	Typ	rhs
min	24	12	16	4	2	3		
ib	0	0	0	0	0	0		
ub	1	1	1	inf	inf	inf		
typ	Con	Con	con	con	con	con		
r_1				1	3		$\geq$	15
r_2				1		2	$\geq$	10
r_3				2	1		$\geq$	20
r_4	-15			1			$\leq$	0
r_5		-20			1		$\leq$	0
r_6			-5			1	$\leq$	0
r_7	4	-3	2				$\leq$	4
r_8	1		1				$\leq$	1
solution	0,6	0,1	0,1	9	2	0,5		58,7

Table 7.2 Implications and new bounds

Setting for probing	Implications and implied bounds
$x_1 = 0$	$\Rightarrow (y_1 \div 0, y_2 \div 20, y_3 \div 5, x_2 \div 1, x_3 \div 1)$
$x_1 = 1$	$\Rightarrow (x_3 \div 0, y_3 \div 0, y_1 \geq 10)$
$x_2 = 0$	$\Rightarrow (y_2 \div 0, y_1 \div 15, x_1 \div 1, x_3 \div 0, y_3 \div 0)$
$x_2 = 1$	No implications and implied bounds
$x_3 = 0$	$\Rightarrow (y_3 \div 0, x_1 \div 1, y_1 \geq 10)$
$x_3 = 1$	$\Rightarrow (x_1 \div 0, y_1 \div 0, y_3 \div 5, y_2 \div 20, x_2 \div 1)$

implemented in MOPS. The models were solved with ClipMOPS which is an 'add-in' to Microsoft Excel and uses the MOPS Dynamic Link Library. Model and solution were directly copied into this document. Table 7.1 contains at the bottom the solution of the corresponding LP-relaxation.

Probing and determination of conditional bounds result in the following implications and new bounds (Table 7.2).

Note that upper bounds of 15, 20 and 5 are derived for the continuous variables y_1, y_2 and y_3.

The extended coefficient reduction is now applied. Consider constraint r_1 transformed into a $\leq$ constraint: $-y_1 - 3y_2 \leq -15$. Let $S = \emptyset$. $U(S)_1 = 0$. Setting $x_1 = 0$ implies $U(S_1^0)_1 = -60$ due to $y_2 = 20$. Since $U(S_1^0)_1 < bu_1$ the coefficient $a_{11} = 0$ can be increased to $a_{11}' = 45$ (after

Table 7.3   The optimal solution

Improved	x_1	x_2	x_3	y_1	y_2	y_3	Typ	rhs
min	24	12	16	4	2	3		
ib	0	0	0	0	0	0		
ub	1	1	1	15	20	5		
typ	bin	bin	bin	con	con	Con		
r_1	45			1	3		$\geq$	60
r_2		5		1		2	$\geq$	15
r_3		10		2	1		$\geq$	30
r_4	-15			1			$\leq$	0
r_5		-20			1		$\leq$	0
r_6			-5			1	$\leq$	0
r_7	1	-1	1				$\leq$	1
r_8	1		1				$\leq$	1
solution	1	1	0	10	1.667	0		79.333

retransformation) and bu_1 can be increased to 60. Similar arguments show that the coefficients a_{22} and a_{32} can be increased from 0 to 5, resp. 10. The coefficient $a_{71} = 4$ and the right-hand-side value bu_7 can be decreased to 2 by simple coefficient reduction (see Johnson *et al.* 1985). We obtain then $U(S_1^0)_7 = 4$. The coefficient $a_{72} = -3$ in constraint r_7 can be increased from -3 to -2. The resulting constraint $2x_1 - 2x_2 + 2x_3 \leq 2$ will be further improved by Euclidean reduction (Hoffman and Padberg, 1991) to $x_1 - x_2 + x_3 \leq 1$. The solution of the resulting LP-relaxation has an integer solution and is, therefore, optimal (Table 7.3).

This type of extended coefficient reduction was critically important to improve the LP-relaxation of the fleet scheduling models as formulated in section II. Most of these models cannot be solved without this type of coefficient reduction. Although the extended coefficient reduction may take a very long time due to many passes many added nonzeros and cuts, this time is easily compensated due to a drastically reduced branch-and-bound-search.

4.4. Node selection strategy

The default node selection strategy was used for solving fleet scheduling problems: Define s_k as the sum of the integer infeasibilities of the LP-solution at node k, i.e.

Table 7.4   Problem data, heuristic, IP-preprocessing, and optimal values

Problem	737a1	737c	320d	737b	310	310b	320h
Flights to be scheduled	293	475	369	579	804	804	887
Model constraints	3119	8975	9661	13063	48250	48406	57303
Model variables	3411	9450	10030	13642	49053	49209	58189
0-1-variables	3119	8975	9661	13063	48250	48406	57303
Model nonzeros	13259	41075	45353	60683	234826	235606	279427
Optimal LP value	66	72	82	59	53	53	157
Value after IP-preproc.	69	73	83	62	54	54	158
Heuristic 1 (LIFO)	79	87	92	90	55	55	168
Heuristic 2 (FIFO)	80	89	89	88	54	62	164
Heuristic 3 (BF)	77	87	90	87	56	59	167
Optimal IP-value	69	73	83	69 (?)	54	54	158

Note: (?) optimal solution unknown.

$$s_k = \sum_j \min(f_{kj}, 1 - f_{kj})$$

where f_{kj} is the fractional part of integer variable j in the LP-solution and the summation is over all integer variables of the problem; thus node k is integer iff $s_k = 0$. Until an integer solution is found we select a node with minimal value of s_k, where ties are broken arbitrarily. After an integer solution was found let z^* be the objective value of the best integer solution found so far and $z(LP)_k$ be the objective function value of the LP-relaxation of node k. A node is selected with a maximal value of $BP_k = (z^* - z(LP)_k)/s_k$.

4.5. Branching variable selection

Each of the three heuristic procedures described in section 2 computes the number of aircraft for a given flight schedule. The heuristic with the smallest solution value zb is used to define the priority vector p for the 0–1-variables. Define $h = 2n + 1$. The priority p_{ij} where j is the immediate successor of flight i (with the same aircraft) is set recursively to $h - 1$, starting with flight 1. Thus, the priority vector is 0 except for the selected successors which are set to decreasing priority values. A branching variable in MOPS is selected as a fractional 0–1-variable with the highest priority.

V. THE OVERALL OPTIMIZATION SYSTEM

A set of tentative flights is given as an ASCII file with the following data: flight number, departure and arrival airports of all legs involved, flight times of all legs, minimum ground time after each leg, fleet to be used, and the departure time window. The data is read into main memory. The heuristic procedures are then executed sequentially. The best one is rerun to produce a priority vector and a tree-bound zb for the branch-and-bound search. The IP-model is then generated directly in main memory arrays of the optimizer. After solving the initial LP and performing the initial supernode processing a general heuristic which is part of the MOPS optimizer is executed. The heuristic is based on rounding and LIFO-enumeration (Suhl, 1985) for 100 nodes. The non-LIFO general purpose branch-and-bound code is executed thereafter. Note that the overall system always produces at least one feasible solution (Figure 7.4).

VI. NUMERICAL RESULTS

The computing environment was a PC with PentiumPro processor (200 Mhz) with 64 MB of main memory. The operating system was Windows 95. All IP-models come from real-life flight schedules of a large European airline. Problem data, solution values of heuristic procedures, objective function values after the initial preprocessing and optimal solution values are shown in Table 7.1.

Table 7.5 shows the total time to apply the initial heuristic procedures, solve the initial LP, perform supernode processing with a tree bound based on the heuristic (one is subtracted because the IP-objective value must be integer), and solve the models with the standard MOPS optimizer where only the branching variable selection is based

Table 7.5   Optimization results with MOPS

Problem	737a1	737c	320d	737b	310	310b	320h
LP-time (mins)	0.03	0.32	0.24	0.58	4.59	4.78	6.53
IP preproc. (mins)	1.34	4.13	10.49	36.21	8.11	28.21	119.97
1. IP-solution value	70	76	83	72	54	54	160
1. IP sol. (nodes)	88	254	322	814	0	154	307
1. IP sol. (mins)	1.49	5.87	14.02	51.19	0.00	33.44	138.43
Best IP-solution value	69	73	83	69	54	54	159
Best IP-sol. (nodes)	842	297	322	914	0	154	319
Best IP-sol. (mins)	3.42	6.11	14.02	52.49	0.00	33.44	138.91
Proving optim. (mins)	3.42	6.11	14.02	480.0*	8.11	33.44	139.89

*Note:** Optimality was not proved in specified time.

on the generated priorities. Optimality was proved in all cases except for problem 737b.

Since the extended coefficient reduction takes a very long time for large models with many 0–1-variables (about 2 hours for model 320h) the question arises if this time is not better used in the branch-and-bound-search. The Table 7.6 shows the impact of initial supernode processing with simple and extended coefficient reduction and the corresponding optimization results. The simple coefficient reduction produces feasible solutions much faster due to drastically reduced IP-preprocessing time. However, finding an optimal solution and proving optimality was not possible in eight hours of computing time. One pass (see Figure 7.3) is defined as the execution sequence of full IP-preprocessing.

Several remarks on the computational results are necessary:

1 The three heuristic procedures are very fast (<1 second) but except for problems 310 and 310b they are not even close to an optimal solution value.

2 Default in MOPS is the simple coefficient reduction. The resulting IP-preprocessing time for the fleet scheduling problems is much shorter compared to preprocessing with the extended reduction.

Table 7.6 Supernode processing and optimization results for problem 320h

	Simple coefficient reduction			*Extended coeff. reduction*		
Cliques, clique entries		1656, 112654				
Implications, Implication entries		3643, 19116				
Implied bounds		8597				
Coefficient modifications	8208			109013**		
Objective value after IP-preproc.	158			158		
CP-time (secs) for IP-preproc.	274.79			7198.26		
No. passes in IP-preprocessing	2			35		
	Value	Nodes	mins	Value	Nodes	mins
First IP-solution	161	513	35.55	160	307	138.43
Best IP-solution	160	530	36.18	158	340	139.89
Proving optimality (mins)		480.0*			139.89	

Note:* Optimality was not proved in specified time.
     ** New nonzero.

Unfortunately in most cases an optimal solution cannot be found and optimality not be proved in 480 minutes computing time without the extended reduction.

3 The time for the extended coefficient reduction for large models could be substantially reduced by a redesign of the probing module. Probing rebuilds the implication at each pass instead of updating it. For problems with many 0–1-variables and many passes in the preprocessing (35 for 320h), efficiency of the implementation is critical.

4 Optimality can be proved only if the optimal solution value is sufficient close to the LP-objective value after the initial IP-preprocessing. If this is not the case (for example, for problem 737b) optimality cannot be proved even after 8 hours. Nevertheless even for such problems the solution values of the heuristic proce-

dures can be substantially improved during the branch-and-bound-process.

5 The difficulties of the models increase if time windows are getting larger, since there are more possible combinations of flights. This situation does not present a serious problem in practice, because there are many rules and regulations reducing the potential size of a time window (see chapter 1).

VII. CONCLUSIONS

Optimal or near-optimal solutions to real-life fleet scheduling problems can be computed in a reasonable amount of computing time. The quality of solutions produced by mixed-integer programming is in general significantly better than the best value produced by three different heuristic procedures. The extended coefficient reduction, together with clique and implication cuts, was critical for solving the models. The heuristic solution procedures were also used to define priorities for the selection of branching variables. This approach reduced the search significantly but alone was not sufficient to solve the models. A substantial reduction of the solution time can be expected by a redesign of probing in conjunction with the extended coefficient reduction in MOPS. From a practical point of view, one would limit the search to a specific CP-time (say, 2 hours) and select the best solution found so far.

References

Baker, E.K., L.D. Bodin and M. Fisher (1995) 'The Development and Implementation of a Heuristic Set Covering Based System for Air Crew Scheduling', *Transportation Policy Decision Making*, 3, 95–110.

Bodin, L., B. Golden, A. Assad and M. Ball (1983) 'Routing and Scheduling of Vehicles and Crews', *Computers and Operations Research*, 10, 63–211.

Brearley, A.L., G. Mitra and H.P. Williams (1975) 'Analysis of Mathematical Programming Problems Prior to Applying the Simplex Algorithm', *Mathematical Programming*, 8, 54–83.

Christofides, N., A. Mingozzi and P. Toth (1979) The Vehicle Routing Problem, *Combinatorial Optimization*, Chapter 11, (New York: Wiley, 1979).

Crowder, H., E.L. Johnson and M.W. Padberg (1983) 'Solving Large-scale Zero–one Linear Programming Problems', *Operations Research*. 31, 803–34.

Desrosiers, J., F. Soumis and M. Desrochers (1984) 'Routing with Time Windows by Column Generation', *Networks*, 14, 545–65.

Desrosiers, J., Y. Dumas, M. Desrochers, F. Soumis, B. Sanso and P. Trudeau (1991) 'A Breakthrough in Airline Crew Scheduling', *GERAD Report*, G-91-11, Montreal.

Dietrich, B.L. and L.F. Escudero (1990) 'Coefficient Reduction for Knapsack Constraints in 0–1 Programs with VUBs', *Operations Research Letters*, 9, 9–14.

Dietrich, B.L., L.F. Escudero and F. Chance (1993) 'Efficient Reformulation for 0–1 Programs – Methods and Computational Results', *Discrete Applied Mathematics*, 42, 147–75.

Gertsbach, J. and Y. Gurevich (1997) 'Constructing an Optimal Fleet for a Transportation Schedule', *Transportation Science*, 11, 20–36.

Guignard, M. and K. Spielberg (1981) 'Logical Reduction Methods in Zero–one Programming (Minimal Preferred Variables)', *Operations Research*, 29, 49–74.

Hoffman, K. and M.W. Padberg (1991) 'Improving LP-representations of Zero–one Linear Programs for Branch-and-cut', *ORSA Journal on Computing*, 3, 121–34.

Johnson, E.L., M.M. Kostreva and U.H. Suhl (1988) 'Solving 0–1 Integer Programming Problems Arising from Large-scale Planning Models', *Operations Research*, 35, 803–19.

Johnson, E.L. (1989) 'Modeling and Strong Linear Programs for Mixed Integer Programming', in S.W. Wallace (ed.), *Algorithms and Model Formulation in Mathematical Programming* (Berlin: Springer Verlag), 1–44.

Levin, A. (1971) 'Scheduling and Fleet Routing Models for Transportation Systems', *Transportation Science*, 5, 232–55.

Savelsbergh, M.W.P. (1994) 'Preprocessing and Probing Techniques for Mixed Integer Programming Problems', *ORSA Journal on Computing*, 6, 445–54.

Suhl, L.M. (1995) *Computer-Aided Scheduling* (Wiesbaden: Gabler Verlag Edition Wissenschaft.

Suhl, U.H. (1985) 'Solving Large-scale Mixed Integer Programs with Fixed Charge Variables', *Mathematical Programming*, 32, 165–82.

Suhl, U.H. (1994) 'MOPS – Mathematical OPtimization System', *European Journal of Operational Research*, 72, 312–22.

Suhl, U.H. and H. Hilbert (1998) A Branch-and-Cut Algorithm for Solving Generalized Multiperiod Steiner Problems in Graphs', *Networks*, 31, 4, 273–82.

Suhl, U.H. and R. Szymanski (1994) 'Supernode Processing of Mixed-Integer Models', *Computational Optimization and Applications*, 3, 317–31.

Wolsey, L.A. (1989) 'Tight Formulations for Mixed Integer Programming: A Survey', *Mathematical Programming*, 45, 173–91.

8 Demand Adaptive Systems: Some Proposals on Flexible Transit[1]

Federico Malucelli, Maddalena Nonato and Stefano Pallottino

I. INTRODUCTION

Traditional public transport systems are evolving towards more flexible organizations in order to capture additional demand and reduce operational costs. Such a trend calls for new models and tools to support the management of new services that make use of public facilities to meet individual need.

A few solutions have already been implemented and are currently available in many urban centers; among them we can mention the *limousine service*, *dial-a-ride service*, and *on-demand door-to-door transportation* for handicapped and elderly people (Ioachim *et al.*, 1995; Psaraftis, 1980; Salvensbergh and Sol, 1995).

Low-demand transportation is a suitable candidate for implementing this kind of service: we speak of low-demand transportation whenever the transportation system is not exploited up to its potential capacity, leaving room for alternative use of the resources involved. Regarding urban transport an example is provided by buses operating during off-peak time slices or during holidays. In all cases the transportation service must be guaranteed, although efficiency is also an issue.

This chapter aims to discuss a new transportation system that provides basic transportation, and at the same time is able to attract additional passengers by allowing the individual user to induce detours in the vehicle routes through a new itinerary closer to the desired one. This system represents a good compromise between an expensive personalized service that precisely fulfills the individual request, such as a taxi ride, and the cheap alternative supplied by the traditional public transport, which may not provide transportation exactly along the requested itinerary.

157

This chapter introduces a new transportation system that we shall call the Demand Adaptive System (DAS), which integrates traditional bus transportation on multiple lines and an on-demand service. The suggested system is designed as follows. Let us consider a set of lines: each traversal of the line is described, as is usual, in terms of a set of timetabled trips. We shall call the stops in the original time table the **compulsory stops**. To introduce some flexibility into the vehicle routes, the vehicle is allowed to transit by each compulsory stop during a **time window**. Beside the compulsory stops, a set of stops to be activated on demand (hereafter, **optional stops**) is available to the users. Between each pair of consecutive compulsory stops a set of optional stops is defined: this is the set of stops that can be visited during the trip from a compulsory stop to the next. Traveling times of the arcs of the physical network are known. A user issues a **service request** specifying a stop where to be picked up and a stop where to be dropped off. In the absence of requests involving optional stops, the vehicle travels along the shortest path on the network within each pair of consecutive compulsory stops. The acceptance of a request implies the rerouting of the vehicle for that part of route involving the optional stop(s) related to the request. Note that the detour may cause a delay of the transit time at the following stops.

The description of the system made above enlightens the differences between a Demand Adaptive System (DAS) and a Demand Responsive System (DRS), such as dial-a-ride, and so on. DAS adapts itself to attract as much demand as possible, but it operates within a conventional line transportation framework. Indeed, users who do not explicitly call for the service, can use DAS in a standard way by boarding and by alighting at compulsory stops only. On the other hand, DRS is an on-demand and personalized service, which usually requires higher costs.

We can distinguish different DAS models depending on the policy used to deal with requests:

DAS1: Requests *may be rejected*, if their acceptance cause infeasibilities, or are not economically worthy. If a request is accepted, users must be picked up and dropped off *exactly* at the stops they asked. Model DAS1 has been introduced in Malucelli *et al.* (1997, 1998).

DAS2: Users are *always* picked up precisely at the requested stop, but they may be dropped off in the vicinity of the requested alighting stop (at the closest compulsory stop, in the worst case), if this cannot be included in the vehicle tour. For this inconvenience the service management pays a penalty to the users, that can be considered in terms of a discount on the travel fare.

DAS3: Users are *always* served, but they can be picked up and/or dropped off at stops in the vicinity of the requested ones, and in these cases the service management pays a penalty.

These models can be further generalised by considering multiple lines and vehicles operating multiple tours along the same line. Starting from the basic model, where a single vehicle runs once along a single line circuit, the model can be generalized by considering vehicles operating multiple tours, multiple interescting lines, and users being allowed to board at one stop of a line and alight at a stop of another line, traveling on vehicles that connect in compulsory stops. In this latter case, synchronization features must be taken into account.

Two alternative policies can be adopted for request scheduling. The first case processes *off-line*. It selects a subset of the requested optional stops such that it maximises the profit and can be feasibly operated. Once the vehicle itinerary has been determined it cannot be modified.

The second scenario processes the requests *on-line*, taking into account the current position of the vehicles with respect to their schedules. The vehicles are rerouted, involving a reoptimization of their schedules and a feasibility check. In this case the problem is highly constrained since previously accepted requests cannot be discarded and scheduled time of boarding stops cannot be anticipated. In this context, time windows act also as a warranty of the quality of the service with respect to requests already accepted. Since rerouting involves the computation of shortest Hamiltonian paths and time response is crucial in the on-line context, exact algorithms are not suitable, while we have to rely upon fast heuristics such as insertion heuristics (Shen *et al.*, 1995) or other approaches that we will introduce later.

In this chapter we tackle the case of a single line and off-line requests' acceptance policy. We formalize the three different service models (DAS1, DAS2 and DAS3) as Mixed-Integer Linear Programming (MILP) problems, analyze their mathematical properties, and suggest heuristic procedures for their solution. Moreover, the models are extended considering a single line with one vehicle operating multiple tours. This study represents the starting point for analyzing the more general case of multiple lines. Finally the on-line problem is briefly discussed.

II. NOTATION AND PROBLEMS DEFINITION

We consider a line structured as a simple circuit, served by a single vehicle, starting and ending its tour at the same terminal. Along the circuit the vehicle passes by a sequence $H = \{f_1, f_2, \ldots, f_{n+1}\}$ of $n + 1$ compulsory stops, where the terminal is the first (f_1) and the last ($f_{n+1} = f_1$) element of the sequence. For each stop f_h a time window $[a_h, b_h]$ is defined; the vehicle must leave f_h not before a_h and not later than b_h, but it may arrive there before a_h for $h = 1, \ldots, n$; b_{n+1} is the maximum trip completion time, and $a_1 = b_1$ is the starting time of the vehicle from the terminal.

A set F_h of optional stops is associated with each pair of consecutive compulsory stops $\langle f_h, f_{h+1} \rangle$. The vehicle passes by an optional stop only if a related boarding or alighting request has been issued. The sets F_h are mutually disjoint. Considering any pair $\langle f_h, f_{h+1} \rangle$ we can define a directed graph $G_h = (N_h, A_h)$, such that $N_h = F_h \cup \{f_h, f_{h+1}\}$ is the stops set and $A_h \subseteq N_h \times N_h$ is the set of arcs connecting the stops. In the following we shall refer to G_h as *segment h*. Finally, $G = (N, A)$ is the whole graph: $G = \cup_h G_h$. The travel time τ_{ij} and the travel cost c_{ij}, for each $(i, j) \in A$ of possible consecutive stops, either compulsory or optional, are given. Without loss of generality, we can suppose that the triangular inequalities hold.

Let us denote by P_h the set of paths in G_h from f_h to f_{h+1}. The vehicle itinerary in segment h is a path $p \in P_h$, having travel time $\tau(p)$ and travel cost $c(p)$ given by the sum of the travel times and the costs of its arcs, respectively:

$$\tau(p) = \sum_{(i,j) \in p} \tau_{ij}; \quad c(p) = \sum_{(i,j) \in p} c_{ij}. \tag{8.1}$$

Let t_h be the starting time from f_h; we assume that $t_1 = a_1$. The sequence of paths defined for each segment forms a *tour q* starting and ending at the terminal. Let us denote by $p_h \in P_h$ the path chosen in segment h; then, the arrival time at the end of the segment, that is at the stop f_{h+1}, is $t_h + \tau(p_h)$. The resulting tour q is *feasible* when:

(i) $t_{h+1} \geq t_h + \tau(p_h) \quad h = 1, \ldots, n-1$

(ii) $a_h \leq t_h \leq b_h \quad h = 2, \ldots, n$

(iii) $t_n + \tau(p_n) \leq b_{n+1}.$

Note that, since no feasible tour can contain a path whose travel time exceeds $b_{h+1} - a_h$, $h = 1, \ldots, n$, we can restrict P_h to the set of paths with travel time less than or equal to $b_{h+1} - a_h$. Let Q be the set of feasible tours; the *global cost* $c(q)$ of tour $q \in Q$ is given by the sum of the costs of the paths forming q.

Let $N(p_h)$, $h = 1, \ldots, n$, be the node set (i.e. the *served stops*) of path $p_h \in P_h$ of segment h, and let $N(q) = \cup_h N(p_h)$ be the set of all served stops.

Let us indicate by R the *request set*; the request $r \in R$ is defined as the pair $\langle s(r), d(r) \rangle$ of boarding and alighting stops; $h(s(r))$ and $h(d(r))$ represent the segments which the boarding stop $s(r)$ and the alighting stop $d(r)$ belong to, respectively. Let us assume that, for each request $r \in R$, $h(s(r)) < h(d(r))$ holds. The assumption that $s(r)$ and $d(r)$ cannot belong to the same segment is quite realistic. Indeed, any two optional stops within the same segment are relatively close to each other. Because of this assumption, no precedence constraints must hold between stops within the same segment, while precedence constraints regarding the pair of stops of each request is implicitly handled by the sequencing of compulsory stops.

As far as model DAS1 is concerned, given a tour q, a request r is *satisfied* only if both the boarding and the alighting stops belong to $N(q)$; the subset $R(q) \subseteq R$ of the requests satisfied by tour $q \in Q$ is given by:

$$R(q) = \{r \in R : s(r), d(r) \in N(q)\}$$

A *benefit* $u(r) \geq 0$ is associated with each request $r \in R$; the *global benefit* $u(q)$ of tour q is given by:

$$u(q) = \sum_{r \in R(q)} u(r)$$

Defining the *profit* as the difference between the benefit and the cost, the problem identified by model DAS1 is to find a tour $q^* \in Q$ of maximum profit:

$$u(q^*) - c(q^*) = \max\{u(q) - c(q) : q \in Q\}$$

As far as models DAS2 and DAS3 are concerned, we notice that all requests are served. However, due to the time window constraints, not all users can be picked up or dropped off (depending on the service model under consideration) at the desired stops. Thus a global penalty measure must be associated with each feasible tour.

Given a request r defined by $\langle s(r), d(r) \rangle$, penalties $v'(r)$ and $v''(r)$ are associated with not serving stop $s(r)$ and $d(r)$, respectively. In this case the user is picked up or dropped off at some other stops in segments $h(s(r))$ and $h(d(r))$, at worst at the closest compulsory stop. Actually, the penalty should depend on which stop is selected for this purpose, therefore depending on the actual vehicle itinerary, but for sake of simplicity, we shall assume the penalty to be proportional to the distance between the missed stop and the closest compulsory stop. That is, the service management offers a discount to the user as s/he where picked up or dropped off at a compulsory stop, even though s/he is allowed to take advantage of any other optional stop included in the actual tour.

Let us introduce the definition of *basic path* of segment h, denoted by $\overline{p}_h = 1, \ldots, n$, as the minimal path with respect to the request management policy. In the case of model DAS2, $\overline{p}_h$ is given by the least travel time path from f_h to f_{h+1} passing by all the optional stops corresponding to boarding requests in segment h, while, in case of model DAS1 and DAS3, the basic path $\overline{p}_h$ is the minimum travel time path from f_h to f_{h+1} without intermediate stops. Note that, in DAS3, $\overline{p}_h$ involves serving the users at the compulsory stops f_h or f_{h+1} rather than at the requested optional stops in segment h, while in model DAS1 $\overline{p}_h$ involves the rejection of all requests concerning optional stops in segment h. Moreover, let $\overline{q}$ denote the basic tour formed by the n basic paths. Note that in DAS2, such a tour might not exist, since even the basic tour passing by all boarding stops can violate the time window constraints. We will discuss this problem in section IV.

As far as models DAS2 and DAS3 are concerned, under the hypothesis of constant penalty made above, each optional stop f not belonging to the basic tour, once introduced in the tour, decreases the penalty by:

$$v(f) = \sum_{r:d(r)=f} v''(r) \tag{8.2}$$

in the case of DAS2, while in the case of DAS3 we have

$$v(f) = \sum_{r:s(r)=f} v'(r) + \sum_{r:d(r)=f} v''(r). \tag{8.3}$$

Given a path $p \in P_h$ we denote by $w(p)$ the *path net worth* with respect to the basic path $\overline{p}_h$:

$$w(p) = \sum_{f \in p, f \notin \bar{p}_h} v(f) - (c(p) - c(\bar{p}_h)) \qquad (8.4)$$

Similarly we can define the *tour net worth* $w(q)$ as the sum of paths net worths:

$$w(q) = \sum_{p \in q} w(p)$$

Thus, regarding model DAS3, let U be the *net benefit* of the basic tour $\bar{q}$ – that is, the difference between the benefit of all requests and the sum of all penalties. While in model DAS2 the net benefit U is given by the difference between the benefit of all requests and the sum of all penalties due to alighting only stops. In models DAS2 and DAS3, the objective is to find the feasible tour q^* that minimizes the global penalty, that is the one maximizing the profit of the basic tour $\bar{q}$ plus the tour net worth of q^*:

$$U - c(\bar{q}) + \max\{w(q) : q \in Q\}.$$

III. DAS1: A REQUEST SELECTION PROBLEM

As previously discussed, the main objective of service model DAS1 is to select the requests to be served, and find a maximum profit feasible tour. The mathematical model makes use of the following variables:

- $z_p^h = 1$ if path $p \in P_h$ is chosen, $z_p^h = 0$ otherwise, $\forall p \in P_h$, $h = 1, \ldots, n$
- $y_r^s = 1$ if $s(r) \in N(p)$, where p is the chosen path ($z_p^h = 1$), $y_r^s = 0$ otherwise, $\forall r \in R$
- $y_r^d = 1$ if $d(r) \in N(p)$, where p is the chosen path ($z_p^h = 1$), $y_r^d = 0$ otherwise, $\forall r \in R$
- t_h = starting time of the vehicle from f_h, $h = 1, \ldots, n$, with $t_1 = a_1$.

Therefore, a request r is served if and only if y_r^s and y_r^d are equal to one.

$$(P1) : \max \sum_{r \in R} u(r) y_r^s - \sum_{h=1}^{n} \sum_{p \in P_h} c(p) z_p^h$$

$$y_r^s \leq \sum_{p \in P_h} \delta_{s(r),p} z_p^h \quad \forall r : s(r) \in N_h, h = 1, \ldots, n \tag{8.5}$$

$$y_r^d \leq \sum_{p \in P_h} \delta_{d(r),p} z_p^h \quad \forall r : d(r) \in N_h, h = 1, \ldots, n \tag{8.6}$$

$$y_r^s = y_r^d \quad \forall r \in R \tag{8.7}$$

$$\sum_{p \in P_h} z_p^h = 1 \quad h = 1, \ldots, n \tag{8.8}$$

$$t_h + \sum_{p \in P_h} \tau(p) z_p^h \leq t_{h+1} \quad h = 1, \ldots, n-1 \tag{8.9}$$

$$t_n + \sum_{p \in P_n} \tau(p) z_p^n \leq b_{n+1} \tag{8.10}$$

$$a_h \leq t_h \leq b_h \quad h = 1, \ldots, n \tag{8.11}$$

$$y_r^s, y_r^d \in \{0, 1\} \quad \forall r \in R$$

$$z_p^h \in \{0, 1\} \quad \forall p \in P_h, h = 1, \ldots, n$$

where $\delta_{s(r),p} = 1$ if $s(r) \in N(p)$, $\delta_{s(r),p} = 0$ otherwise, $\forall p \in P_{h(s(r))}$, $\forall r \in R$, and $\delta_{d(r),p} = 1$ if $d(r) \in N(p)$, $\delta_{d(r),p} = 0$ otherwise, $\forall_p \in P_{h(s(r))}$, $\forall r \in R$.

Notice that variable y_r^s is equal to one iff the vehicle passes by stop $s(r)$ and request r is served, and, similarly, y_r^d is equal to one iff the vehicle passes by stop $d(r)$ and request r is served.

Constraints (8.5) and (8.6) link the choice of the path to the served requests; constraints (8.7) couple boarding and alighting stops for each request. Constraints (8.8) impose the selection of one path for each segment, while constraints (8.9), (8.10) and (8.11) state the requirement that the selected paths form a feasible tour.

Problem $(P1)$ is NP-Hard since a particular instance reduces to a TSP. Later on, we will discuss some methods to compute upper bounds and heuristic solutions for the problem.

3.1. DAS1+: a multiple tour service model

Let us now consider a service model where, instead of having one vehicle operating a single tour along the circuit line, K tours $\{q_1, \ldots,$

$q_K\}$ of the same circuit line are performed. The multiple tours can be performed by a single vehicle consecutively, or by multiple vehicles. For the sake of simplicity, we focus on the case of a single vehicle, though the model can be straightforwardly generalized to the case of multiple vehicles. In the multiple tour case, a request $r \in R$ is specified by a triplet $\langle s(r), d(r), i(r)\rangle$ where $i(r)$ is the *ideal service tour*. Instead of simply discarding a request, the service management can decide to serve it in a tour different from the ideal one. Thus, the problem is actually a requests assignment to tours. In the following we give a mathematical formulation of the problem. Let $[a_h^i, b_h^i]$ be the time window at compulsory stop f_h during the ith tour, $h = 1, \ldots, n, i = 1, \ldots, K$. Since the tours are sequentially operated, we can assume $b_{n+1}^i \le a_1^{i+1}$ for $i = 1, \ldots, K - 1$.

It is reasonable to assume that the benefit of a request depends on the tour it is assigned to: being zero if the request is served too late or too early with respect to $i(r)$, and decreasing as the service delay or earliness increases. Let $u^i(r)$ be the benefit associated with request r if served during the ith tour, for each $r \in R, i = 1, \ldots, K$. Note that, conversely to the case of the single tour, a request $r = \langle s(r), d(r), i(r)\rangle$ can have $h(d(r)) < h(s(r))$ – that is, a user can alight in a stop which is located earlier in the line with respect to the boarding stop. In this case the user is picked up during one tour and is dropped off during the successive tour. Let R' be the set of requests such that $h(s(r)) < h(d(r))$, and $R'' = R/R'$.

The mathematical model is the following:

$$(P1+): \max \sum_{i=1}^K \sum_{r \in R} u^i(r)y_r^{s,i} - \sum_{i=1}^K \sum_{h=1}^n \sum_{p \in P_h} c(p)z_p^{h,i}$$

$$y_r^{s,i} \le \sum_{p \in P_h} \delta_{s(r),p} z_p^{h,i} \quad \forall r: s(r) \in N_h, h = 1, \ldots, n, i = 1, \ldots, K \quad (8.12)$$

$$y_r^{d,i} \le \sum_{p \in P_h} \delta_{d(r),p} z_p^{h,i} \quad \forall r: d(r) \in N_h, h = 1, \ldots, n, i = 1, \ldots, K \quad (8.13)$$

$$y_r^{s,i} = y_r^{d,i} \quad \forall r \in R', i = 1, \ldots, K \quad (8.14)$$

$$y_r^{s,i} = y_r^{d,i+1} \quad \forall r \in R'', i = 1, \ldots, K - 1 \quad (8.15)$$

$$\sum_{i=1}^K y_r^{s,i} \le 1, \quad \forall r \in R \quad (8.16)$$

$$\sum_{p \in P_h} z_p^{h,i} = 1 \quad h = 1, \ldots, n, i = 1, \ldots, K \tag{8.17}$$

$$t_h^i + \sum_{p \in P_h} \tau(p) z_p^{h,i} \le t_{h+1}^i \quad h = 1, \ldots, n-1, i = 1, \ldots, K \tag{8.18}$$

$$t_n^i + \sum_{p \in Pn} \tau(p) z_p^{n,i} \le b_{n+1}^i \quad i = 1, \ldots, K-1 \tag{8.19}$$

$$a_h^i \le t_h^i \le b_h^i \quad h = 1, \ldots, n, i = 1, \ldots, K \tag{8.20}$$

$$y_r^{s,i}, y_r^{d,i} \in \{0, 1\} \quad \forall r \in R, i = 1, \ldots, K$$

$$z_p^{h,i} \in \{0, 1\} \quad \forall p \in P_h, h = 1, \ldots, n, i = 1, \ldots, K$$

where variables t_h^i give the starting time from stop f_h in the ith tour, variables $y_r^{s,i}, y_r^{d,i}$ are equal to one if the request r is served in the ith tour, and variable $z_p^{h,i}$ is equal to one if path $p \in P_h$ belongs to the ith tour. Constraints (3.16) state that all requests must be assigned to a tour at most.

If we consider the case where more vehicles operate on the same line, constraints (3.15) become:

$$y_r^{s,i} = y_r^{d,i+V} \quad \forall r \in R'', i = 1, \ldots, K-V \tag{8.21}$$

where V is the number of vehicles operating on the line. Obviously, as far as time windows are concerned, $b_{n+1}^i \le a_1^{i+V}$, for $i = 1, \ldots, K - V$.

The solution approach to this problem can be similar to the one adopted for the single-tour case, even though the size is much larger and other decompositions could be introduced.

IV. DAS2: MINIMIZING THE PENALTY OF ALIGHTING STOPS

Provided that all requests are served and all users are picked up at the desired stops, the main objective of service model DAS2 can be seen as defining a maximum net worth feasible tour. For the sake of simplicity we assume that there exists a basic tour $\bar{q}$, that is a tour passing by all boarding stops and fulfilling the time window requirements. Note that, in this model, P_h is the set of all paths in G_h passing by at least all nodes corresponding to boarding stops of segment h and fulfilling the time.

Let U be the net benefit of $\bar{q}$, that is:

$$U = \sum_{r \in R} u(r) - \sum_{r:d(r) \notin \bar{q}} v''(r).$$

where $v''(r)$ is the alighting penalty associated with request r. Let $\overline{F}_h$ be the set of optional stops not in the basic path $\overline{p}_h$ given by the path passing by all boarding stops of segment h:

$$\overline{F}_h = \{f \in F_h \setminus N_h(\overline{p}_h)\}$$

For each optional stop $f \in \overline{F}_h$ we can compute the saving with respect to U resulting from the insertion of f in the path, as in (8.2), and for each feasible path $p \in P_h$ we can compute the *net worth* with respect to the basic path $\overline{p}_h$ as stated by (8.4).

Note that the net worth $w(\overline{p}_h)$ of the basic path $\overline{p}_h$ is equal to zero. The mathematical model of the problem of finding the feasible tour maximizing the net worth with respect to $\overline{q}$ is the following:

$$(P2): \max \sum_{h=1}^{n} \sum_{p \in P_h} w(p) z_p^h$$

$$\sum_{p \in P_h} z_p^h = 1, \quad h = 1, \ldots, n \tag{8.22}$$

$$t_h + \sum_{p \in P_h} \tau(p) z_p^h \leq t_{h+1}, \quad h = 1, \ldots, n-1 \tag{8.23}$$

$$t_n + \sum_{p \in P_h} \tau(p) z_p^n \leq b_{n+1} \tag{8.24}$$

$$a_h \leq t_h \leq b_h, \quad h = 1, \ldots, n \tag{8.25}$$

$$z_p^h \in \{0, 1\} \quad \forall p \in P_h, h = 1, \ldots, n$$

Constraints (8.23), (8.24), and (8.25) state the feasibility of the departure times: the vehicle must leave f_h after it has arrived, and within the time window. It should be remarked that problem $(P2)$ has a block structure, where each block corresponds to a segment. Note that the proposed model involves variables z_p^h only, while variables y_r^s and y_r^d have been omitted with respect to model $(P1)$. This is due to the fact that all requests are served and the selection of the stops to be served is implicit in the path choice and in the definition of $w(p)$. The simple formulation suggests a solution approach based on Column Generation methods, as discussed in Sub-section 6.2.

4.1. DAS2+: a multiple tour service model

As mentioned in the previous sections, the basic tour that passes by all boarding stops can be infeasible – that is, it may violate the time-windows constraints in some compulsory stops. In this circumstance, the service management may decide to discard some requests, reducing the problem to the one seen in the case of DAS1. Alternatively, we can think of a service system with multiple vehicles, or with a single vehicle making multiple tours. In this case, instead of discarding requests, the service management has to decide in which tour a request has to be served. As in DAS1+, a different benefit can be associated with each request, depending on the tour in which it is served. As in sub-section 3.1, let $z_p^{h,i}$ be a variable equal to 1 when path p of segment h is selected during the ith tour, and 0 otherwise, and let $y_r^{s,i}$ be a variable saying if request r is served during the ith tour, that is the vehicle passes by stop $s(r)$ during the ith tour. Moreover, $y_r^{d,i}$ is equal to one if the vehicle passes by stop $d(r)$ in the ith tour and request r is served. The mathematical model is:

$$(P2+): \max \sum_{i=1}^{k} \sum_{r \in R} u^i(r) y_r^{s,i} - \sum_{r \in R} v_r'' \left(1 - \sum_{i=1}^{k} y_r^{d,i}\right) +$$

$$- \sum_{i=1}^{K} \sum_{h=1}^{n} \sum_{p \in P_h} c(p) z_p^{h,i}$$

$$y_r^{s,i} \le \sum_{p \in P_h} \delta_{s(r),p} z_p^{h,i} \quad \forall r : s(r) \in N_h, h=1,\ldots,n, i=1,\ldots,K \quad (8.26)$$

$$y_r^{d,i} \le \sum_{p \in P_h} \delta_{d(r),p} z_p^{h,i} \quad \forall r : d(r) \in N_h, h=1,\ldots,n, i=1,\ldots,K \quad (8.27)$$

$$y_r^{s,i} \ge y_r^{d,i} \quad \forall r \in R', i=1,\ldots,K \quad (8.28)$$

$$y_r^{s,i} \ge y_r^{d,i+1} \quad \forall r \in R'', i=1,\ldots,K-1 \quad (8.29)$$

$$\sum_{i=1}^{K} y_r^{s,i} = 1, \quad \forall r \in R \quad (8.30)$$

$$\sum_{i=1}^{K} y_r^{d,i} \le 1, \quad \forall r \in R \quad (8.31)$$

$$\sum_{p \in P_h} z_p^{h,i} = 1 \quad h = 1, \ldots, n, i = 1, \ldots, K \tag{8.32}$$

$$t_h^i + \sum_{p \in P_h} \tau(p) z_p^{h,i} \le t_{h+1}^i \quad h = 1, \ldots, n-1, i = 1, \ldots, K \tag{8.33}$$

$$t_n^i + \sum_{p \in Pn} \tau(p) z_p^{n,i} \le b_{n+1}^i \quad i = 1, \ldots, K-1 \tag{8.34}$$

$$a_h^i \le t_h^i \le b_h^i \quad h = 1, \ldots, n, i = 1, \ldots, K \tag{8.35}$$

$$y_r^{s,i}, y_r^{d,i} \in \{0,1\} \quad \forall r \in R, i = 1, \ldots, K$$

$$z_p^{h,i} \in \{0,1\} \quad \forall p \in P_h, h = 1, \ldots, n, i = 1, \ldots, K$$

Note that constraints (8.31) are redundant being implied by constraints (8.28), (8.29), and (8.30).

The assignment of requests to tours is explicitly specified by way of variables $y_r^{s,i}$. In the objective function the contribution of the benefit of a request r depends on the tour the request is assigned to (stated by variables $y_r^{s,i}$). Moreover, the penalty in case of displacement (depending on variables $y_r^{d,i}$) and the cost of the selected paths have to be subtracted. Similarly to DAS1+, the model can be generalized in order to deal with multiple vehicles.

V. DAS3: MINIMIZING THE PENALTY OF BOARDING AND ALIGHTING STOPS

In service model DAS3 all requests are served, but users, instead of being picked up and/or dropped off at the desired stops, can be picked up and/or dropped off at alternative stops. This opportunity allows the service management to determine a feasible tour in any demand condition. As in the case of DAS2 the inconveniences caused to the users will be payed in terms of penalties (discounts on the transit fare). The approach is similar to that of DAS2, though simplified. Let us redefine some concepts taking into account the additional degree of freedom introduced in the boarding stops. For each segment h the basic path $\overline{p}_h$ goes straight from f_h to f_{h+1}, thus the set of optional stops to be considered is:

$$\overline{F}_h = \{f \in F_h : \exists r \in R, f = d(r) \text{ or } f = s(r)\}$$

For each optional stop $f \in \bar{F}_h$ we can compute the saving with respect to U as stated by (8.3); as in the case of DAS2, for each feasible path $p \in P_h$ we can compute the *net worth* with respect to the basic path $\bar{p}_h$ as in (8.4).

The mathematical model of the problem of finding the feasible tour maximizing the net worth with respect to $\bar{q}$ is the following, and is exactly the same of (*P2*) except for the definition of $w(p)$, $\bar{F}_h$, and P_h. In particular, the set of feasible paths P_h is, as in DAS1, the set of paths going from f_h to f_{h+1} and it includes the one of the models DAS2, since paths are not obliged to pass by all boarding stops of segment h.

$$(P3): \max \sum_{h}^{n} \sum_{p \in P_h} w(p) z_p^h$$

$$\sum_{p \in P_h} z_p^h = 1, \quad h = 1, \ldots, n \tag{8.36}$$

$$t_h + \sum_{p \in P_h} \tau(p) z_p^h \leq t_{h+1}, \quad h = 1, \ldots, n-1 \tag{8.37}$$

$$t_n + \sum_{p \in P_n} \tau(p) z_p^n \leq b_{n+1} \tag{8.38}$$

$$a_h \leq t_h \leq b_h, \quad h = 1, \ldots, n \tag{8.39}$$

$$z_p^h \in \{0, 1\} \quad \forall p \in P_h, h = 1, \ldots, n$$

Note that in the case of DAS3 there always exists a feasible solution, for example $\bar{q}$.

5.1. DAS3+: a multiple tour service model

As in the case of DAS2, model DAS3 can be extended to comply with multiple tours. Note that, by contrast with DAS2+, even if a request is boarded during the ith tour, both $y_r^{d,i}$ and $y_r^{s,i}$ are equal to 0 whenever the user is not boarded and alighted at the requested stops. Therefore, a new set of variables is needed to denote the requests assignment to tours: let y_r^i be equal to 1 if request r is boarded during the ith tour, and 0 otherwise. The mathematical model is the following:

$$(P3+): \max \sum_{i=1}^{K} \sum_{r \in R} u^i(r) y_r^i - \sum_{r \in R} v_r' \left(1 - \sum_{i=1}^{k} y_r^{s,i}\right)$$

$$- \sum_{r \in R} v_r'' \left(1 - \sum_{i=1}^{k} y_r^{d,i}\right) - \sum_{i=1}^{K} \sum_{h=1}^{n} \sum_{p \in P_h} c(p) z_p^{h,i}$$

$$y_r^{s,i} \leq \sum_{p \in P_h} \delta_{s(r),p} z_p^{h,i} \quad \forall r : s(r) \in N_h, h = 1, \ldots, n, i = 1, \ldots, K \quad (8.40)$$

$$y_r^{d,i} \leq \sum_{p \in P_h} \delta_{d(r),p} z_p^{h,i} \quad \forall r : d(r) \in N_h, h = 1, \ldots, n, i = 1, \ldots, K \quad (8.41)$$

$$\sum_{i=1}^{K} y_r^{s,i} \leq 1, \quad \forall r \in R \tag{8.42}$$

$$\sum_{i=1}^{K} y_r^{d,i} \leq 1, \quad \forall r \in R \tag{8.43}$$

$$\sum_{i=1}^{K} y_r^i = 1, \quad \forall r \in R \tag{8.44}$$

$$y_r^{s,i} \leq y_r^i \quad \forall r \in R, i = 1, \ldots, K \tag{8.45}$$

$$y_r^{d,i} \leq y_r^i \quad \forall r \in R', i = 1, \ldots, K \tag{8.46}$$

$$y_r^{d,i+1} \leq y_r^i \quad \forall r \in R'', i = 1, \ldots, K-1 \tag{8.47}$$

$$\sum_{p \in P_h} z_p^{h,i} = 1 \quad h = 1, \ldots, n, i = 1, \ldots, K \tag{8.48}$$

$$t_h^i + \sum_{p \in P_h} \tau(p) z_p^{h,i} \leq t_{h+1}^i \quad h = 1, \ldots, n-1, i = 1, \ldots, K \tag{8.49}$$

$$t_n^i + \sum_{p \in Ph} \tau(p) z_p^{n,i} \leq b_{n+1}^i \quad i = 1, \ldots, K-1 \tag{8.50}$$

$$a_h^i \leq t_h^i \leq b_h^i \quad h = 1, \ldots, n, i = 1, \ldots, K \tag{8.51}$$

$$y_r^{s,i}, y_r^{d,i}, y_r^i \in \{0, 1\} \quad \forall r \in R, i = 1, \ldots, K$$

$$z_p^{h,i} \in \{0, 1\} \quad \forall p \in P_h, h = 1, \ldots, n, i = 1, \ldots, K$$

for the sake of clarity, we explicitly formulate constraints (8.42) and (8.43), even though they are redundant, since implied by (8.44), (8.45), (8.46), and (8.47).

As for DAS1+ and DAS2+, model DAS3+ can be generalized to deal with multiple vehicles.

Model DAS3+ can be extended in order to guarantee a good service level as, for example, by introducing the rule according to which a passenger either is displaced in space or in time. In the first case a request r is not served at the requested stops, but it must be served during the ideal tour $i(r)$; in the other case a request is not served during tour $i(r)$, but it must be served at the requested stops. This condition is enforced by the following set of constraints:

$$\sum_{i \neq i(r)} y_r^{s,i} + y_r^{i(r)} = 1 \quad \forall r \in R, i = 1, \ldots, K \tag{8.52}$$

$$\sum_{i \neq i(r)} y_r^{d,i} + y_r^{i(r)} = 1 \quad \forall r \in R', i = 1, \ldots, K \tag{8.53}$$

$$\sum_{i \neq i(r)} y_r^{d,i+1} + y_r^{i(r)} = 1 \quad \forall r \in R'', i = 1, \ldots, K-1 \tag{8.54}$$

Note that the same rule can be introduced in model DAS2+ also, by adding constraints (8.53) and (8.54) where $y_r^{i(r)}$ is replaced by $y_r^{s,i(r)}$.

VI. SOLUTION APPROACHES

In this section we propose three possible approaches to the models described so far. The first is a Lagrangian Relaxation, which is applicable to ($P1$) but it can be extended to ($P1+$), ($P2+$), and ($P3+$). The second approach is a Column Generation, which is suitable for ($P2$), and ($P3$), where request selection variables are not present. These approaches, beside an upper-bound value, yield other information which can be exploited in the construction of a heuristic solution. A possible heuristic approach exploiting this information is discussed, too.

6.1. Lagrangian-relaxation

Let us illustrate the case of DAS1.

The Lagrangian-relaxation of constraints (8.7), (8.9), and (8.10) with multipliers λ and μ, respectively, yields a set of separable sub-problems, one for each segment h, of the form:

$$(P(h)_{\lambda,\mu}):\max \sum_{r:s(r)\in N_h}(u(r)-\lambda_r)y_r^s + \sum_{r:d(r)\in N_h}\lambda_r y_r^d$$

$$-(\mu_h-\mu_{h-1})t_h - \sum_{p\in P_h}(c(p)+\mu_h\tau(p))z_p^h$$

$$y_r^s \le \sum_{p\in P_h}\delta_{s(r),p}z_p^h \quad \forall r:s(r)\in N_h$$

$$y_r^d \le \sum_{p\in P_h}\delta_{d(r),p}z_p^h \quad \forall r:d(r)\in N_h$$

$$\sum_{p\in P_h}z_p^h=1$$

$$a_h \le t_h \le b_h$$

$$y_r^s \in \{0,1\} \quad \forall r:s(r)\in N_h$$

$$y_r^d \in \{0,1\} \quad \forall r:d(r)\in N_h$$

$$z_p^h \in \{0,1\} \quad \forall p\in P_h$$

Each sub-problem $(P(h)_{\lambda,\mu})$, beside being of smaller size with respect to the global problem $(P1)$, is also quite affordable to solve. In fact the request benefits (including the effect of multipliers λ) can be included in a node weight of graph G_h, and the cost of the links (including the effect of multipliers μ) can be included in an arc weight of the same graph. Thus the problem is a maximum-path problem with a constraint on the travel time. It should be noted that in graph G_h, with the above weight definitions, positive cycles may occur; however, due to the time-windows constraints, the problem is bounded and can be efficiently solved, as described in Desrosiers *et al.* (1995).

The Lagrangian Dual is solved by way of a bundle algorithm (Carraresi *et al.*, 1996), which provides an upper bound to the problem, a collection of feasible paths for each segment, and a set of optimal multipliers.

6.2. A Column Generation approach

Problems $(P2)$ and $(P3)$ can be naturally approached through Column Generation. Let us briefly describe the method in the general case, since the algorithm can be easily specialized considering the different definitions of P_h and $w(p)$ given for DAS2 and DAS3. The master problem $(\overline{P})$ is the LP relaxation of $(P2)$ or $(P3)$ which considers paths in $P'_h \subseteq P_h, h = 1, \ldots, n$, only. According to the Column Generation method, $(\overline{P})$ is solved to optimality. Then, paths in $P_h \backslash P'_h$ are searched such that, if considered, would improve the current solution. Initially P'_h may contain $p \, \overline{p}_h$ only, or a set of feasible paths heuristically determined. Due to the block structure, the search of a path to enter the problem can be decomposed into subproblems, one for each segment. The master problem $(\overline{P})$ is the following:

$$(\overline{P}): \max \sum_{h=1}^{n} \sum_{p \in P'_h} w(p) z_p^h$$

$$\sum_{p \in P'_h} z_p^h = 1 \quad h = 1, \ldots, n \tag{8.55}$$

$$t_h + \sum_{p \in P'_h} \tau(p) z_p^h \leq t_{h+1} \quad h = 1, \ldots, n-1 \tag{8.56}$$

$$t_n + \sum_{p \in P'_h} \tau(p) z_p^n \leq b_{n+1} \tag{8.57}$$

$$a_h \leq t_h \leq b_h, \quad h = 1, \ldots, n$$

$$0 \leq z_p^h \quad \forall p \in P'_h, h = 1, \ldots, n$$

Let π_h, σ_h be the optimal dual variables of problem $(\overline{P})$ associated with constraints (8.55), (8.56), and (8.57), respectively. The Column-Generation phase seeks a feasible path in any segment whose reduced cost with respect to π_h, σ_h is greater than zero. If such a path p exists for some h, then the current solution of $(\overline{P})$ can be improved by adding p to P'_h; when such a path does not exist for each h, the current solution is optimal for the LP relaxation of $(P2)$ or $(P3)$. Let us briefly analyze the problem of searching a positive reduced cost path in segment h.

The reduced cost of a path $p \in P_h$ is:

$$\overline{w}(p) = w(p) - \pi_h - \sigma_h \tau(p), \quad \forall p \in P_h, h = 1, \ldots, n$$

by applying the definition of path net worth (8.4), path travel time, and path cost (8.1), we get:

$$\overline{w}(p) = -\pi_h - \sum_{(i,j) \in p} (c_{ij} + \sigma_h \tau_{ij}) + \sum_{f \in \overline{F}_h \cap p} v(f) + c(\overline{p}_h)$$

The search for a feasible path with positive reduced cost can be accomplished by looking for the longest feasible path from f_h to f_{h+1} in G_h, where arcs have weight $-(c_{ij} + \sigma_h \tau_{ij})$ and nodes have weight $v(f)$ if $f \in \overline{F}_h$, and zero otherwise. In the case of DAS2, the path has to pass by all the boarding nodes, while for DAS3 this restriction does not hold. This problem is easily solved by means of longest path algorithms; actually, paths may contain positive cycles, but the problem reduces to a classical maximum length path on a suitable space–time network (Pallottino and Scutellà, 1998).

6.3. Heuristic approaches

Since all the problems considered so far can be viewed as a particular *Vehicle Routing Problem*, classical heuristics are suitable to approach them. In particular, we can devise several 'insertion' heuristics (Shen *et al.*, 1995), where a tour is built from the basic one by iteratively inserting new optional stops in order to satisfy more requests or reduce penalties. Insertion criteria may vary depending on the transportation model.

Beside the aforementioned heuristics, another kind of appoach can be conceived. The proposed algorithm exploits the particular structure of the problem and the information yielded by the Lagrangian Relaxation or the Column Generation. In particular both methods yield a set of 'promising' paths. However, such a set can be suitably integrated by other methods. The algorithm assembles paths in a greedy fashion, selecting them from the set of promising paths, trying to build a maximum profit tour.

The approach, with minor modifications, is suitable for all the three models so far introduced (DAS1, DAS3, and DAS3), and can be easily adapted to deal with the multiple tour cases.

The algorithm can be summarized as follows. Consider a set of paths $P' = \cup_h P'_h$, where P'_h corresponds to the set of promising paths of segment h. Let the basic tour $\overline{q}$ made by the n basic paths $\overline{p}_h$ be the starting solution. The algorithm iteratively improves the value of the current feasible tour q by swapping basic paths in q with paths in P' according to a greedy criterion. At each step, the *most promising* path with respect to the current tour q is selected and removed from P'; the meaning of most promising path depends on the kind of transportation model, as we will see later. Let h be the segment of such a path, then the algorithm *checks the feasibility* of the new tour obtained from q by swapping the selected path with $\bar{p}_h$. If feasible, P' is updated by removing all paths in P'_h, so that the selected path will belong to the final solution. The algorithm stops when P' becomes empty.

Two issues must be addressed in order to implement this procedure: how to check the feasibility of the tour yielded by the swapping, and the criteria according to which paths are selected.

Regarding the feasibility check of the new tour q, note that the path relative to segment h, denoted by $p_h(q)$, can be thoroughly characterized by the following information: the earliest departure time (EDT), the latest departure time (LDT), the earliest arrival time (EAT), and the latest arrival time (LAT), which can be iteratively computed according to:

$$EDT(p_h(q)) = \begin{cases} a_1 & \text{if } h = 1 \\ \max\{a_h, EDT(p_{h-1}(q)) + \tau(p_{h-1}(q))\} & \text{if } h = 2, \ldots, n \end{cases}$$

$$LDT(p_h(q)) = \begin{cases} \min\{b_n, b_{n+1} - \tau(p_n(q))\}; & \text{if } h = n \\ \min\{b_h, LDT(p_{h+1}(q)) - \tau(p_h(q))\} & \text{if } h = n-1, \ldots, 1 \end{cases}$$

$$EAT(p_h(q)) = EDT(p_h(q)) + \tau(p_n(q)) \quad h = 1, \ldots, n \tag{8.58}$$

$$LAT(p_n(q)) = LDT(p_h(q)) + \tau(p_h(q)) \quad h = 1, \ldots, n \tag{8.59}$$

Let τ_h be the maximum travel time for paths in P'_h in order to qualify for swapping.

$$\tau_h = LDT(p_{h+1}(q)) - \max\{EAT(p_{h-1}(q)), a_h\}$$

Therefore, a path $p \in P'_h$ can be swapped with $\bar{p}_h$ only if $\tau(p) \leq \tau_h$.

Regarding the path selection criteria, paths in P' may be ranked according to a score $s(p)$ that is a measure of the benefits coming from including path p in the final solution, with respect to the current solution q.

Relatively to models DAS2 and DAS3, the contribution of path p is given exactly by the *net worth* $w(p)$ which, by definition, gives the gain obtained by swapping the basic path with p. Note that the score does not depend on the other $n - 1$ paths of the tour.

On the other hand, in the case of model DAS1, the score of a path does depend on the other paths in the tour, in particular the benefit of a request whose alighting node is in p contributes to the value of the solution only provided that the tour passes by the boarding node (and vice versa). Therefore, unlike the previous case, the score of path p varies according to the current solution q.

Let us restate the profit of tour q in terms of pairs of paths. Let $\theta(p, p')$, with p and p' in different segments, be the benefit of all requests having the boarding node in p and the alighting node in p' or vice versa. Then

$$u(q) - c(q) = \sum_{h=1}^{n-1} \sum_{h'=h+1}^{n} \theta(p_h, p'_h) - \sum_{h=1}^{n} c(p_h(q)) \qquad (8.60)$$

but, with respect to the current solution q, only the paths resulting from a swap are known to belong to the final solution.

Given $p \in P'_h$ the score $s(p)$ can be computed as follows. For each segment $l \neq h$ for which a path has not been already selected, the path p_l such that $\tau(p_l) \leq \tau_l$ and which maximizes $\theta(p, p_l) - c(p_l)$ is considered. The score $s(p)$ corresponds to the profit of the tour given by the selected paths, path p, and all paths p_l according to (8.60).

Note that, after any successful swapping of $\bar{p}_h$ with $p \in P'_h$, $EDT(p'_h(q))$ must be updated for all $h' \geq h$, $LDT(p'_h(q))$ must be updated for all $h' \leq h$, while EAT and LAT are recomputed according to (8.58) and (8.59), respectively. As a consequence, for each segment h for which a path has not been already selected, τ_h may have been decreased, thus reducing the cardinality of set P'_h. In case of model DAS1 more updating is necessary since some paths p_l may not belong to P'_h any more, and the scores of some nodes may have to change.

VII. ON-LINE APPROACHES

The ideal and more realistic service in low-demand conditions is an *on-line* service, where requests can arrive also during the service duty of the vehicle. We may assume that some basic rules characterize an on-line service:

- once a request has been accepted, it cannot be *discarded*
- once a service has been determined for a request (pick up stop, drop off stop, tour which it has been assigned to), it cannot be *modified* or in general *worsened*

- the service request reservations must be processed in a *first-come-first-serve* fashion
- the service request reservation must be confirmed (i.e. accepted, rejected or modified proposing alternative pick up and/or drop off stops, and tour of assignment) in *real time*.

Actually, these rules simplify the optimization problem that must be solved each time a request is issued, since the number of feasible alternatives to the current solution is much smaller with respect to the off-line case. Let us briefly discuss how the solution approaches to DAS1, DAS2 and DAS3 change under the on-line perspective. Assume that the current established feasible tour is q and a new service request r arrives asking to be served starting from the current instant of time, and the vehicle has not passed by the desired boarding stop.

In the DAS1 framework, the problem is to decide whether $s(r)$ and $d(r)$ can be included in q, if not already present, maintaining the feasibility. There are two possible approaches:

- recomputing a new optimal tour q' including $s(r)$ and $d(r)$, and if a feasible solution is found, r is accepted and q' is the new current tour; the request is rejected, otherwise
- heuristically evaluating if $s(r)$ and $d(r)$ can be inserted into q: if the answer is negative, the request is discarded even though there is not theoretical evidence that no feasible tour including r exists; if the answer is positive, the current tour can be refined after having confirmed the acceptance of the request.

Another possible way to implement a real-time answer to service requests is to store a set of good feasible solutions in a database and efficiently retrieve the best among the available ones.

In the DAS2 case, the problem is slightly more delicate as it has to be decided whether the insertion of $s(r)$ into q is feasible, and whether it is more profitable to insert $d(r)$ into q or to pay a penalty $v''(r)$. Also in this case there are two possible approaches. The exact one computes the optimal tour, the heuristical one evaluates the feasibility of the possible insertion of $s(r)$ and the feasibility/profitability of the possible insertion of $d(r)$. In the heuristical approach the evaluation can be followed by a refining phase where the optimal tour is recomputed exactly, or with more accuracy.

In the DAS3 case, the problem is simply to evaluate the profitability of making a detour for boarding and/or alighting request *r*. This can be done by applying the same ideas seen in the case of DAS2.

The DAS1+, DAS2+, and DAS3+ systems in the on-line case can be approached very simply. Provided that it is not reasonable to serve a request with more than one tour of delay with respect to the ideal one, it is sufficient to *unroll* the circuit line twice – that is, the planning horizon is of two circuits instead of one. This unrolling can be dynamically updated as segments are visited by the vehicle.

7.1. Hybrid on-line solutions

The on-line approach requires a quite heavy technological support: vehicle monitoring systems, communication system between the service management center and the vehicles, and between the stops and the service management center. Often, in a preliminary testing phase, a cheaper system with fewer technological requirements is highly desirable, even though it can not implement all the features of the on-line service described above. Let us consider a possible hybrid on-line system. In such a system, requests can arrive while the vehicles are operating, but the itinerary of each vehicle is determined on the basis of the data available at the beginning of the tour, and cannot be modified while the vehicle is running. In practice, the service management has to solve a sequence of single-tour problems, ((*P*1) or (*P*2) or (*P*3), according to the system selected) just before the vehicle leaves the terminal. The problems are solved using the service requests collected while the vehicle is performing the current tour and those that could not be previously served. In order to avoid a request being indefinitely held, a suitable priority mechanism has to be introduced, such that in the solution of a single-tour problem older requests have better chances to be served.

VIII. CONCLUSIONS AND FUTURE WORK

In this chapter we presented a new transit model identified as Demand Adaptive System (DAS). The system can be seen as a hybrid solution between a conventional line transit and an on-demand personalized transportation. This solution should combine the advantages of the two systems: the low costs and the reliability of a line transit on one side, and the flexibility and the capability of attracting users of a personalized transportation service on the other side. In the DAS framework we proposed several transportation models that are characterized by the different ways of managing the service requests. For each model we proposed a mathematical formulation and solution techniques in the case of single-line systems.

A transportation system like the one proposed has several interesting aspects, especially nowadays. From the transportation companies' point of view, the DAS can be suitable for improving the efficiency and for maintaining (or even enhancing) the service level; this point appears more and more critical. From the technological point of view, the tools needed to monitor the vehicles in the network, the communication systems between vehicles, users, and management center, as well as an efficient and clear information system for the users are challenging subjects. Indeed many of these systems have been already implemented and adopted by many transit companies.

From the mathematical and algorithmical viewpoint the proposed model shed a new light on how to approach a transportation problem. We believe that this kind of transportation systems deserve to be studied and tested in practice in the near future.

The preliminary computational results regarding the off line models (Malucelli *et al.*, 1998) are encouraging. Some test problems have been generated in order to reproduce possible urban settings with up to 10 segments, a number of stops ranging from 25 to 140, and a number of requests ranging from 40 to 190. The proposed relaxation computes a solution in a reasonable time (order of minutes of a medium-sized workstation), and its quality is acceptable even for large instances. The value of the relaxation favorably compares with that of the trivial linear programming relaxation obtained by relaxing the integrality constraints and by solving the problem by means of standard LP codes. In fact, the linear-relaxation is usually quite poor, while the Lagrangean-Relaxation is always more tight. The heuristic algorithms proposed produced the optimal solution in a very short time for the small instances, where it has been possible to compute the optimal solution with a branch and bound. For larger instances the value of the heuristic solution is not far from the upper bound given by the relaxation. The implementation of all the proposed algorithms does not require any particular computational platform (an up-to-date PC is sufficient) nor any particular commercial software.

Model extensions should deal with a system with multiple DAS lines. When the lines share some stops (both compulsory and optional) two levels of problem arise. One passenger may ask to be transported from an optional stop of one line to an optional stop of another line. First of all, the system has to decide where to allow the passenger to transfer from one line to another. Moreover, it is necessary to enforce the synchronization of the vehicles involved by the transfers. All those aspects introduce mathematical difficulties, especially when the off-line planning is considered and will be the subject of further work.

Note

1. This work has been supported by grant no. 97.000256.PF74, National Research Council.

References

Carraresi P., A. Frangioni and M. Nonato (1996) 'Applying Bundle Methods to Optimization of Polyhedral Functions: An Applications Oriented Development', *Ricerca Operativa*, 25, 5–49.

Desrosiers, J., Y. Dumas, M.M. Solomon and F. Soumis (1995) 'Time Constrained Routing and Scheduling', in M.O. Ball *et al.* (eds), *Network Routing, Handbooks in Operations Research and Management Science*, 8 (Amsterdam: Elsevier Science), 35–139.

Ioachim I, J. Desrosiers, Y. Dumas, M.M. Solomon and D. Villeneuve (1995) 'A Request Clustering for Door to Door Handyicapped Transportation', *Transportation Science*, 29, 63–78.

Malucelli F., M. Nonato and S. Pallottino (1997) 'Modelli di Trasporto Collettivo in Condizioni di Domanda Debole', *Atti del III Convegno del Progetto Finalizzato Trasporti*, 2, Taormina.

Malucelli F., M. Nonato and S. Pallottino (1998) 'Models and Heuristic Algorithms for Low Demand Transportation Systems', preprints of TRISTAN III (Puerto Rico).

Pallottino, S. and M.G. Scutellà (1998) 'Shortest Path Algorithms in Transportation Models: Classical and Innovative Aspect', in P. Marcotte and S. Nguyer (eds), Proceedings of the International *Colloquium on Equilibrium in Transportation Models* (Dorarecut: Kluwer), 245–81.

Psaraftis H.N. (1980) 'A Dynamic Programming Solution to the Single Vehicle Many to Many Immediate Request Dial-a-ride Problem', *Transportation Science*, 14, 130–54.

Salvensbergh, M.W.P. and M. Sol (1995) 'The General Pickup and Delivery Problem', *Transportation Science*, 29, 17–29.

Shen Y., J.Y. Potvin, J.-M. Rousseau and S. Roy (1995) 'A Computer Assistant for Vehicle Dispatching with Learning Capabilities', *Annals of Operations Research*, 61, 189–211.

9 A Scheduling Prototype for Factory Automation: Matching OR Methodologies to Actual Industrial Needs

Roberto Tadei, Federico Della Croce and Salvatore Pucci

I. INTRODUCTION

At the end of the twentieth century, the goal of any planning/scheduling expert is still the generation of a package able to fully integrate customers orders to the final production Gantt chart.

A belief widely held by industrial managers and even many Operations Research (OR) practitioners is that the classical scheduling theory (as part of Operations Research) deals with fictitious problems that are still too far from those faced in industry to be of use in practical scheduling problems. Indeed, in industrial applications, the classical scheduling theory appears inadequate even in its semantic phase. When transferred to the real industrial environment, the links between customers' and suppliers' orders, bills of materials, working cycles and inventories, on the one hand, and machines, jobs, processing times, release and due dates on the other are often quite obscure.

Currently, for both safety and fiscal reasons, most medium and small-size European factories have their inventory data and customers' and suppliers' orders stored on mainframes. For this reason, most of the current administrative packages running under mainframes cover inventory management and consequently material requirement planning, but not the production scheduling aspect. Moreover, most of these factories look for packages that use the administrative output and the availability of production resources inputs to provide the final production scheduling, typically represented by a Gantt chart. Our work is located in the above context and it attempts to fill the gap

183

between the real industrial needs and the theoretical scheduling approach. We present an applied scheduling prototype, which uses administrative and production data to derive the required outputs. The scheduler runs on a PC and is based on Operations Research methodologies. Before the installation the scheduling process has been successfully validated and verified in several case studies (see Tadei *et al.*, 1995, for example). The present chapter presents the test results concerning an Italian factory that produces paper packaging and has a job shop production structure.

II. APPLICATION TERMINOLOGY

An initial on-site analysis highlighted the fact that the necessary input data for the planning/scheduling process were available in soft format only on mainframe, where all the accounting and other administrative matters were handled. Many medium/small-sized factories adopt this practice as for confidentiality reasons the data relating to the factory accounting (e.g. suppliers' orders) cannot be handled on a PC. For the above reason, many accounting packages nowadays also offer as sub-routine a Material Requirement Planning (MRP) program in order to handle the planning of suppliers' orders on the same workstation used for the accounting. We stuck to this situation and focused only on the scheduling aspect assuming that raw materials' input data are available. Nonetheless there was still quite a difference between the data we needed as input for the scheduling package and the data supplied by the factory management system. This is due to the actual discrepancy in the terminology used by the industrial community compared to the classical scheduling theory. The first step was therefore the creation of a small dictionary, which matched industrial terms with academic ones. We report below the main terms used in industry and their correspondence with the scheduling terminology.

- Customer order
  The customer order is the **main document in the production flow**. Any other production document exists only as a consequence of a customer order. A customer order is denoted by the item to be produced and by the required quantity. The full set of customer orders denotes the total factory workload while the required production quantities concur to generate the production order

once the inventory of given items has been considered and a lot-sizing step has been accomplished. The customer order does not have a specific correspondence in classical scheduling theory terminology, which takes into account directly the production orders (jobs).

- Production order
  The **production order** in the industrial environment corresponds to the **job** in classical scheduling theory: it is denoted by a release time defined by the arrival of raw material and intermediate products and a due date which corresponds to the required delivery date for that production order. The production order is denoted also by the item to be produced and by the required quantity: these two components define the set of job operations through the bill of materials and the working cycle.
- Bill of materials
  For each item there exists a bill of materials that defines the **exact requirements of raw materials and intermediate products**.
- Working cycle and cycle phases
  For each item there exists a **working cycle**: this cycle is split into a sequence of **phases**. Each phase is denoted by the working center (centers) to which that phase is assigned and by the kind of working process to be accomplished. Each phase is also denoted by the time required to process a single unit of product.
- Supplier order
  A supplier order is generated (and the corresponding supplier alerted) when a determined raw material which is necessary to satisfy a production order is **not available in sufficient quantity in the storehouse**. It is characterized by a delivery date, which is a constraint to the starting time of the working cycles requiring that raw material. This date concurs (together with the delivery dates of all other raw materials necessary for the processing of a product order) to define the release time of the job as it is defined in classical scheduling theory. In our case, we did not have to deal with the suppliers' orders which were considered by the planning system, but only with the release time of the corresponding job.
- Working center
  The working center is the unit **dedicated to the processing of a phase of the working cycle**. It is generally a single machine or a set of unrelated/uniform/parallel machines. It matches exactly the definition of machine (or set of unrelated/uniform/parallel machines) applied in classical scheduling theory.

- Work order
  A set of work orders is associated with each production order, with
  an order for each working cycle phase. It represents the **production
  lot to be processed on the given working center**. The equivalent in
  classical scheduling theory is the operation of a given job.

III. A REAL SCHEDULING PROBLEM

The factory concerned is located in North-West Italy and produces
paper packaging. The case study considers a production of boxboard
divided into two main types:

- the first consists of boxes made of folding boxboard (the type used
  for foodstuffs such as pasta, rice, sugar, salt, etc.)
- the second consists of boxes made of corrugated board (the type
  used for the display of cosmetics or for small white electrical
  goods).

The production structure of the firm is organized in work 'islands',
each containing a working center composed of a machine, a small
work-in-process (WIP) stock (production buffer) and a store of tools
(materials to be consumed, punches, etc.).

The factory has on average 4/5 alternative working centers available
for each working phase, but some phases can be undertaken only
at a single working center, which frequently becomes a bottleneck. The
number of customer orders dealt with is around 1000 per year.
In general, there is a one-to-one correspondence between customer
orders and production orders. Particularly large customer orders,
however, are divided into several production orders with the same
delivery date. In all, it is possible that a total of approximately 2000
production orders could be processed per year. The company mainly
aims to fulfil the delivery dates. Minimizing WIP and maximizing working time of each working center are secondary objectives.

3.1. Components' list and products' working cycle

Folding boxboard type

The list of components consists of a single element, the folding
boxboard (known as 'liner') which comes in two forms: White/Grey or

White/White according to the colors of the two faces. It can also differ in thickness. The type of liner used depends solely on the finished product. The liner is stored in the warehouse in rolls of different length (from a minimum of 20–30 m to a maximum of 1000 m) but a single width (120 cm).

The working cycle includes the following phases:

- collection of the required liner from the store
- liner printing (the rolls are printed in 'shapes' such that each shape can contain a certain number of boxes and has a standard length, equal to the roll width)
- cutting of liner into sheets
- covering of liner sheets with plastic film
- die cutting of the sheets to obtain single boxes
- window formation (application of plastic film in certain areas covering holes to allow foodstuff to be visible through the 'window')
- formation of box
- packaging and storage in warehouse.

Each of these phases requires specific resources and/or machines. Not all the phases above are always necessary, however, as some finished products do not require plastic film covering or 'windows'.

Corrugated board type

The list of components consists of two elements: the liner (as in the previous type) and the corrugated medium named *flute*. The liner is always White/Grey. The flute can be of various kinds: different thickness and different densities. Liner and flute have rolls of various lengths.

The working cycle includes the following phases:

- collection of the required liner from the store
- liner printing (*see* the folding boxboard working cycle)
- collection of the required flute from the store
- liner/corrugated medium splicing
- cutting of the flute into sheets
- covering of flute sheets with plastic film
- die cutting of the sheets to obtain single boxes
- window formation

- formation of box
- packaging and storage in warehouse.

For each type of product, the total production and the required materials are driven by the demand, plus an extra margin to allow for wastage due to testing (printing, die cutting, etc.). The timing of each working phase is calculated according to the speed of the machine in full production and in the transition phases at the beginning and end of each job. Each phase can be carried out in different working centers with different outputs (new machines can be 5–6 times as fast as the old ones for the same job). The time required for each phase is therefore a function of the working center, in which the phase is carried out.

The factory contains 20 working centers supporting several phases of working cycles. Some working centers carry out particular phases required by specific jobs only (e.g. window formation). Other phases are required by all jobs (e.g. cutting into sheets). Each working center can be active between 1 and 3 8-hour shifts a day. The production is made up of 80 different types of products, each of which involves some or all of the phases listed above.

IV. OR APPROACH TO THE SCHEDULING PROBLEM

The presence of different types of product with different working cycles implies different job routings in the scheduling problem. This in turn implies a job-shop-like structure (Lawler *et al.*, 1993). As different machines can process the same working cycle phase of a given job with different processing times, we also have a context of unrelated parallel machines. Hence our problem is a generalized job shop problem where a single operation can be processed on a set of unrelated parallel machines (while in a classical job shop model it can be processed on just one machine). We therefore have to solve both a sequencing problem and an assignment problem. The main goal of the factory is to meet the due dates: more precisely the management looks for an appropriately weighted *minimization* of:

- the total *number of jobs* that violate the due date
- and the *maximum* of these violations.

This corresponds, in classical scheduling terminology, to the *minimization* of total number of tardy jobs and *maximum* lateness.

Formally the objective function of the problem is:

$$\min \beta N_\tau + \alpha L_{max}$$

with:

$$N_\tau = \sum_{i=1,\ldots,n} X_i$$

$$L_{max} = \max_{i=1,\ldots,n} C_i - d_i$$

$$\alpha, \beta \in [0,1]: \alpha + \beta = 1$$

where:

α and β are the constant weights of each objective
n = number of jobs
d_i = due date of order i
C_i = completion time of order i
$X_i = 0$ if $C_i - d_i \leq 0$, otherwise $X_i = 1$

Notice that every time there exists a solution with $L_{max} \leq 0$, then both objectives are met as all the due dates are satisfied, while if $L_{max} > 0$, at least one due date is violated and also $N_\tau > 0$.

To keep a similar notation to the 3-field classification of Lawler *et al.* (1993), we can denote the above problem as $J_R \| \beta N_\tau + \alpha L_{max}$, where J_R indicates that we are dealing with a job shop (J) problem where each operation can be processed on a set of unrelated parallel machines (R).

4.1. The solution methodology

Production scheduling problems belong to the area of *combinatorial optimization*. Combinatorial optimization is related to problems where the number of solutions is finite. This could suggest the use of explicit enumeration, namely the generation and evaluation of all feasible solutions. Such an approach, however, is intractable for relatively simple problems as the number of feasible solutions grows exponentially with the number of jobs. Furthermore for many combinatorial optimization problems (and this is even more true for scheduling problems) there currently exists no exact algorithm able to find an optimal solution within limited time. The possibility of finding such an algorithm is considered pretty unrealistic. To make it short (for a more comprehensive discussion on this subject, we refer to Lawler *et al.*, 1993), these problems are denoted as *NP-Hard* and the only practical way to deal with them is to apply heuristic solution procedures which do not guarantee obtaining the optimal solution, but require an acceptable CPU time. The basic job shop problem ($J \| C_{max}$) is well known to be *NP-Hard* (Lawler *et al.*, 1993). This implies that our problem too is *NP-Hard* as it generalizes that problem. In our case, we propose a heuristic solution procedure based on the idea of local (neighborhood) search (Reeves, 1993). More precisely we propose a *Steepest Descent Local* search procedure based on the Tabu search (Glover, 1989, 1990) concept.

Let X be the solutions set and $x \in X$ a generic feasible solution. We denote with $V(x)$ the set of those solutions that can be derived from x by applying some properly defined modification. This modification is such that, in general, any solution $\hat{x} \in V(x)$ is *similar* to x. The set $V(x)$ is the so-called *neighborhood* of x.

A classical steepest descent local search method consists of the following steps:

1 Initialization:
  – Find a starting solution $x_1 \in X$
  – Consider it as the current solution $\bar{x}$ ($\bar{x} = x_1$)
2 Neighborhood generation:
  – Generate the neighborhood $V(\bar{x})$ of the current solution $\bar{x}$
  – Evaluate each element $\hat{x} \in V(\bar{x})$
  – Let $x^* \in V(\bar{x})$ be the best solution found.
3 Acceptance test:
  – IF x^* is better than $\bar{x}$, THEN
  – set $\bar{x} = x^*$
  – GO TO 2
  – ELSE GO TO 4
4 Stop.

It should be noted that in a classical local search procedure the solution quality is strongly biased by the neighborhood structure. This procedure has the disadvantage of having no way to escape from local optima. This fact is illustrated in Figure 9.1, where all the solutions belonging to the neighborhood $V(x_n)$ are worse than x_n. However, a little further away there is the global minimum, which cannot be reached by the local search procedure once it gets trapped in x_n.

Tabu Search procedure (TS) is a so-called *metaheuristics* derived

Figure 9.1 Local and global minimum in a local search procedure

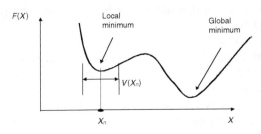

from the steepest descent local search with some further features that help to avoid being trapped in local minima. Given the current solution, its neighborhood is evaluated, the best solution x^* is derived and a deterministic decision on the acceptance of x^* is taken. In this case, however, x^* is candidate to become the new current solution even if it does not improve upon $\bar{x}$. If the neighborhood structure is symmetric, namely if $\bar{x} \in V(x^*)$, by setting x^* as new current solution a loop could occur if $\bar{x}$ is the best solution in $V(x^*)$. That depends on the fact that the following iteration would go back to $\bar{x}$.

To avoid this loop (and more general ones) the move which modifies the old current solution $\bar{x}$ into the new current solution is recorded in a list called *tabu list* and kept there for a predefined number of iterations. Every time a neighborhood is evaluated, its best solution is accepted only if the moves leading to that solution are not present in the tabu list. An exception to this rule is allowed only when the best solution, though obtained through a *tabu move* (a move belonging to the tabu list), improves upon the best solution even reached. This is a so-called *aspiration criterion*. Finally a stopping test is added as the classical local search stopping criterion no longer holds.

A general tabu search procedure is composed of the following steps:

1 Initialization:
- Find a starting solution $x_1 \in X$
- set $\bar{x} = x_1$
- empty the tabu list TL: $TL = \{\}$
- reset the iterations counter I: $I = 0$.

2 Derive $V(\bar{x})$ from $\bar{x}$ and evaluate $V(\bar{x})$ or a sub-set $V'(\bar{x}) \subset V(\bar{x})$, let x^* be the best solution found such that:

x^* satisfies an aspiration criterion OR

the move $m(\bar{x}, x^*)$ that modifies $\bar{x}$ into x^* is not tabu ($m(\bar{x}, x^*) \notin TL$), THEN

- set $\bar{x} = x^*$
- insert $m(x^*, \bar{x})$ in the tabu list ($TL = TL \cup m(x^*, \bar{x})$)
- if the tabu list has reached its maximum size, delete from TL the oldest move
- increase the iterations counter ($I = I + 1$).

3   Execute the stopping test; if it is positive, STOP; ELSE GO TO 2.

4   STOP.

In order to design a tabu search procedure, the following decisions need to be taken:

- definition of the neighborhood structure and decision on the set of neighbors to visit at each iteration: the whole neighborhood or just a sub-set and, in the latter case, the size of this subset
- definition of the kind of moves to be inserted in the tabu list
- definition of the length of the tabu list
- definition of the aspiration criterion/criteria
- definition of a stopping test, for instance the total number of iterations.

4.2. The implemented solution procedure

The implemented solution splits in three sequential steps:

- search for a starting solution by means of a list scheduler (corresponding to step 1 of the above tabu search procedure)
- improvement of the starting solution by means of a local search method working on a sub-set of the defined neighborhood
- search for the final solution by means of a local search method working on the whole neighborhood.

Search for a starting solution

The starting solution is derived by means of a list scheduler, which schedules the jobs operations one after another in the following way:

1   **Each job is given a priority index.** This index is computed by subtracting the sum of the processing times of all the job operations to the job due date. We remark that, as each operation identifies the

corresponding working center but not the machine to which it will be assigned. In order to compute the processing time, the fastest machine of that working center is considered. This index represents the so-called *lead time* of that job – namely, the latest starting time which may allow a completion time not greater than the due date.

2 **The jobs are sorted in increasing order of lead time**, forming a priority queue.

3 **The first job** j of the priority queue is considered and the working center on which its first unprocessed operation is scheduled is evaluated.

4 Given the working center, the job operation is temporarily assigned to the machine M_i on which it will be the **first to be completed** and its temporary starting time S_{j,M_i} defined.

5 Now **all the other jobs ready to be processed on** M_i are considered. If there exists a job $k \neq j$ ready to be processed on M_i and to be completed before S_{j,M_i}, then job k is scheduled on M_i and the algorithm goes to step 7.

6 Otherwise job j is **scheduled on** M_i.

7 **The priority index of the jobs is updated**. If a job has completed its working cycle, it is deleted from the priority queue. If the priority queue is empty the algorithm stops, otherwise it goes back to step 3.

The above procedure finds a solution, which takes into account the capacity constraints of the machines but may violate some of the due dates. This solution is the starting solution to which the local search procedure is applied. If we denote with n the number of jobs and with m the number of machines, the computational complexity of this phase is $O(mn^2)$.

Improvement of the starting solution

The aim of this second phase is to improve the starting solution by applying a tabu search approach with the constraint of using very limited time. This implies that at each tabu search iteration only a restricted number of neighbors can be evaluated. In order to define a suitable representation of the problem so as to come up with an appropriate neighborhood, the following considerations need to be made.

With a job shop type of problem, given a feasible schedule, it is well known that an arbitrary swap of two operations on a machine may lead to an unfeasible schedule (French, 1982). Hence, the representation of the problem as a set of strings (one for each working center) does not allow interchanges in the operations processed on that center that still guarantee feasible schedules.

For our problem, we chose to apply the approach proposed in Della Croce (1995) for the job shop problem that applies the Generalized Pairwise Interchange (GPI) operators as neighborhood operators.

The GPI operators are:

- API (**Adjacent Pairwise Interchange**): a switch between two adjacent jobs. Given the sequence 'a–b–**c**–**d**–e–f–g–h', by swapping **c** and **d** we obtain 'a–b–**d**–**c**–e–f–g–h'
- NAPI (**Nonadjacent Pairwise Interchange**): a switch between two nonadjacent jobs. Given the sequence 'a–b–**c**–d–e–f–**g**–h', by swapping jobs **c** and **g** we obtain 'a–b–**g**–d–e–f–**c**–h'
- EFSR (**Extraction and Forward Shifted Reinsertion**). A switch of two nonadjacent jobs according to the following scheme. Given the sequence 'a–b–**c**–d–e–f–**g**–h', job c is extracted from its position and reinserted immediately after job **g**. The resulting sequence is: 'a–b–d–e–f–**g**–**c**–h'
- EBSR (**Extraction and Backward Shifted Reinsertion**): the reverse of the previous switch. Given the sequence 'a–b–**c**–d–e–f–**g**–h', job g is extracted from its position and reinserted immediately before job **c**. The resulting sequence is: 'a–b–**g**–**c**–d–e–f–h'.

The GPI neighborhood in the job shop problem can be described as follows. Consider a job shop problem with 4 jobs ($J1, J2, J3$ and $J4$) and 3 machines (M_a, M_b and M_c). The routing of job $J1$ is M_a (4)–M_b (5)–M_c (3), while that of $J2$ is M_b (2)–M_c (6)–M_a (1), that of $J3$ is M_c (3)–M_b (3)–M_a (5) and that of $J4$ is M_a (4)–M_c (2)–M_b (3), where the figures in brackets are the corresponding processing times. Consider now an aggregate sequence with $4 \times 3 = 12$ elements, one for each operation. The aggregate sequence will therefore have three elements referring to each job $J_i(i = 1, \ldots 4)$, respectively. Suppose, for instance, the aggregate sequence is

$$J1 \ J4 \ J2 \ J2 \ J3 \ J1 \ J3 \ J4 \ J4 \ J3 \ J1 \ J2$$

this sequence can be interpreted as the relative importance of the operations, where the operations of each job keep the order required by the job routing. In the example proposed, as $J1$ is the starting element, the first operation of $J1$ (which is on machine M_a and whose length is 4) is scheduled as soon as possible. The first operation of *J*4 (which is also on machine M_a) is scheduled next, starting immediately after the first operation of *J*1. In the same way, the first operation of *J*2 is scheduled to start at time 0 on machine M_b, the second operation of *J*2 at time 2 on machine M_c, and so on. In this manner each aggregate sequence corresponds to a feasible solution and all GPI operators can be properly applied. Notice that the API and NAPI operators become meaningless if the elements *i,j* refer to the same job. To extend this approach to the generalized job shop, given the working center, it is sufficient to choose the machine on which the chosen job completes first. In this phase, we evaluate at each iteration a sub-set of the above neighborhood, which is obtained by considering only swaps among operations, processed on the same working center. Every time a swap is accepted, the jobs involved in the swap are added to the tabu list. The length of the tabu list is equal to the number of working centers. The aspiration criterion adopted here allows a tabu move if the new solution improves upon the best solution ever reached. This neighborhood is quite restricted, but guarantees experimentally obtaining a good solution in a short time.

Search for the final solution

The third phase applies to the whole neighborhood listed above. Hence, swaps of operations processed on different working center are now also considered. Any two operations working on different working centers can be exchanged, provided that there exists in between at least one operation which is processed on either of these working centers. This further requirement is applied in order to avoid evaluating solutions coinciding with the current one.

Stopping test

Two stopping criteria are considered. The first involves a maximum number of consecutive iterations with no improvement. The user can, however, select also a time limit stopping criterion useful for large size instances.

V. COMPUTATIONAL RESULTS

The modified tabu search procedure has been implemented using the C++ programming language. The tests have been performed on a Sun Ultra-Enterprise workstation. We applied it to both real and randomly generated problem instances. In all instances, the first stopping criterion has been applied. Given the complexity of the problem (a generalization of the job-shop problem for which it is already difficult to compute good lower bounds) and the type of objective function involved it was not possible to derive in the real instances meaningful lower bounds to validate the quality of the solution.

Hence, as far as those instances are concerned, we provide a table (Table 9.1) indicating the CPU times in seconds of the list scheduler (LS), the first local search (L_1) and the second local search (L_2), respectively and the relative improvement of the second local search vs. the first local search $(LS - L_1)/(LS - L_2)$. This was done in order to estimate the general behavior of the procedure on the factory instances. Table 9.1 shows that the first local search always provides good solutions that are very near to the final solutions obtained with the second local search. In addition, the first local search procedure requires much less CPU time than the second. For the large size instances the second local search was not able to improve the solution obtained by the first local search. As a result of these considerations, we can expect that the larger the instances, the bigger is the CPU time gap between the first and the second local search procedures.

To validate the procedure in terms of quality of the solution, we used the Muth–Thompson instances (1963) (whose optimal solution is well known) for the $J \| C_{max}$ problem which is a special case of our problem. Table 9.2 depicts the performance of the procedure. We run the

Table 9.1   Real problems

Problem	Jobs	Working centers	LS CPU	L_1 CPU	L_2 CPU	$\dfrac{LS - L_1}{LS - L_2}$ Sol. value
TDP20	20	30	0.1	1.1	3.1	0.86
TDP30	30	30	0.1	6.5	17.6	0.96
TDP40	40	30	0.1	21.6	59.8	0.97
TDP50	50	30	0.1	51.2	166.2	1.00
TDP60	60	30	0.1	94.3	301.1	1.00
TDP70	70	30	0.1	143.4	501.4	1.00
TDP100	100	30	0.2	477.7	1881.6	1.00

Table 9.2   Muth–Thompson problems

Problem	Jobs	Machines	LS Avg. sol.	LS CPU	L_1 Avg. sol.	L_1 CPU	L_2 Avg. sol.	L_2 CPU	L_2 Best sol.	Opt. sol.
MT06	6	6	59.8	0.1	56.3	0.4	55	2.0	55	55
MT10	10	10	1181.3	0.1	1042.0	5.9	988.5	58.2	938	930
MT20	20	5	1593.7	0.1	1285.9	12.1	1230.6	62.3	1189	978

procedure 100 times for each instance introducing some uncertainty in the list scheduler so as to start the local search from 100 different schedules. From the results we notice a good-quality performance of the procedure even though it was not specifically tailored for the C_{max} objective function.

VI. CONCLUSIONS

The purpose of this chapter is to attempt to bridge the current gap between classical scheduling theory and practical industrial scheduling problems. The three-step procedure implemented based on Tabu Search gives promising results in terms of CPU time requirements and solution quality when tested on randomly generated and real-world instances. From that experience we expect future enhancement of industrial automation based on OR methodologies. The synergetic effect of direct and heuristic optimization techniques could be taken as a paradigm for the development of new management tools in the factory automation of the twenty-first century.

References

Della Croce, F. (1995) 'Generalized Pairwise Interchanges and Machine Scheduling', *European Journal of Operational Research*, 73, 310–19.

French, S. (1982) *Sequencing and Scheduling: an Introduction to the Mathematics of the Job-Shop* (Chichester: Horwood).

Glover, F. (1989) 'Tabu Search, Part I', *Orsa Journal on Computing*, 1, 190–206.

Glover, F. (1990) 'Tabu Search, Part II', *Orsa Journal on Computing*, 2, 4–32.

Lawler, E.L., J.K. Lenstra, A.H.G. Rinnooy Kan and D.B. Shmoys (1993) 'Sequencing and Scheduling: Algorithms and Complexity', *Handbooks in Operations Research and Management Science*, Vol. 4: *Logistics of Production and Inventory* (Amsterdam: North-Holland), 445–524.

Muth, J.F. and G.L. Thompson (1963) *Industrial Scheduling* (Englewoods Cliff, NJ: Prentice-Hall).

Reeves, C.R. (ed.) (1993) *Modern Heuristic Techniques for Combinatorial Problems* (Oxford: Blackwell Scientific).
Tadei, R., M. Trubian, J.L. Avendaño, F. Della Croce and G. Menga (1995) 'Aggregate Planning and Scheduling in the Food Industry: A Case Study', *European Journal of Operational Research*, 87, 564–73.

10 Supply-chain Production Planning

Javad Ahmadi, Robert Benson
and Daniel Supernaw-Issen

I. INTRODUCTION

In this chapter we describe development of a production planning system for Advanced Micro Devices (AMD), a major supplier of integrated circuits. This development is motivated by AMD's efforts in development of a Total Order Management System for higher quality and faster response to customer orders. The Order Management System contains a statement of supply availability for a foreseeable future. Based on this statement of availability, new orders are assessed and responded to. Associated with the Order Management System is a production planning system that takes as input the statement of demand and produces an optimized solution plan. The planning process is regenerative and executed on a regular basis to update the statement of supply via updated demand and production status data. There is room in the Order Management System to execute exception-based net-change scenarios when particular important order requests do not match the availability outlook. Our focus in this chapter is the presentation of the planning system.

The semiconductor production process from raw silicon start to finished packages is a fairly extensive process that requires many major stages of processing over a geographically dispersed network of production facilities owned by AMD or its vendor base of manufacturing service providers. The overall cycle time for production of a finished product may take well over several months requiring up to 500 discrete manufacturing steps. This lengthy and complicated production process is further complicated by the large variety of products under production and mixed volumes of demand from low sporadic production of certain products to high-volume, high-frequency production. For such lengthy production processes the planning horizon is naturally very lengthy, adding to the complexity and size of the planning models. Well over 10 000 products may be actively involved in the supply chain at any time. At a high-level view, the major production stages consists of:

- Wafer fabrication, which starts with raw silicon wafers, upon which in a repetitive set of processes multi-layer micron dimension circuitry is created. This is the most time-consuming process owing to the large number of discrete and batch process steps required. While the number of production centers in this stage are few, the repetitive nature of the wafer fabrication process, characterized as re-entrant flows, creates a complicated production environment due to set-up customization requirements, finite capacity, and volatile yields. Anywhere where from 100 to 500 steps may be required to create a finished wafer.
- Wafer sort, which identifies the good die or grade of die on a finished wafer, since a wafer consists of many dies. The sort process involves a few major steps and is typically bounded by capacity of the testing machines and testing fixtures. The overall cycle time at this stage may be less than a few days, which is also typical of the following production stages.
- Die assembly, which starts by cutting out the good dies from the wafer and encapsulation in various packaging materials such as plastics or ceramics with interconnection form choices.
- Chip test, which tests and grades packaged chips. This is also known as the bining process since a given die may result into various performance grades.
- Mark and pack, which completes the process by marking and packaging the product into standard forms such as tubes, reels or boxes with supporting materials, ready for distribution.

In this industry the product structure is an expanding graph where a few hundred finished wafers become several thousand finished products. However, the graph is not a tree due to bining and substitutions. Substitution allows for more that one manufacturing product to be used as parent product for the next manufacturing product. Product production history – trace ability – is very important in this industry as production processes are under continuous improvement and learning cycles. All production occurs in discrete lots and lots are tracked for execution and performance history. In conjunction with multi-sourcing of production arising from multiplicity of facilities, there is a huge burden on the nomenclature process to provide the capability to identify each product by stage of completion, location of completion or method of execution. Such a product produced at one facility may be

Figure 10.1   Product structure graph

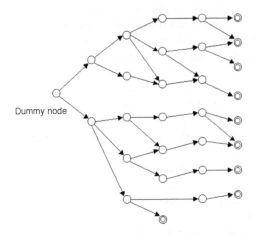

Dummy node

named differently if produced at a different facility. In fact, certain customer restrictions require manufacture of a product to occur at specific facilities. In this manner, the true product structure is at best a directed graph with a hierarchical appearance with respect to the production stages as shown in Figure 10.1. Not all products follow the restricted production stages hierarchy, certain facilities are capable of completing several stages together. In case of multi-chip module production, near-finished products (marked products) are used as starting materials in assembly and may continue into the following stages. Purchased materials, in particular in assembly areas, play a significant role and require representation in planning process.

II. THE PLANNING MODEL

Our objective is to develop an enterprise-wide global planning system that is capacity and materials feasible with respect to time-varying production cycle times and yields, encompassing a host of production controls such as inventory levels, with full accounting of the work-in-process while maximizing service level, revenue, and minimizing cost. The challenge from the modeling perspective is the management of problem scale and hence solution time for utility of the tool as an affordable periodic regenerative decision support tool.

The core problem is a well known production/inventory linear programming model encompassing a set of planning periods and decisions variables covering production decisions, inventory levels related to a set of periodic product inventory balance equations, backlog measures, capacity, and materials limits. The most complicated aspect of the model is timely relation of production decisions and resources requirements. Each product production decision creates a sequence of process activities based on the product process flow. Each process activity requires a set of resources at a particular level, and has a cycle time. Modeling at the activity level requires scheduling decisions' representation. Manufacturing activities have cycle times of minutes or hours, which makes activity level representation within an extended planning horizon of nearly a year impossible. In all implementations of such planning systems modeling is done at product production stage levels – i.e. inter-stage planning, while inner-stage capacity requirements are represented by implications. Therefore associated with each product production decision at a given stage, a schedule of resource demands is used to relate the effects of the decision. In other words, modeling is done at the inter-stage requisite resources. This method requires assumptions in terms of how activities occur on a production route in terms of inter-step queueing delays, process times, yields, etc. For this purpose, nominal data is used. Simulation models of the factory floor provide estimates of such data, and essentially provide a *picture* of the implications. If a production process route in a stage is too long for one aggregate or macro representation, the route may be broken into segments for a more flexible representation at the expense of additional variables and equations. Additionally, alternative implication pictures may be supplied to compensate for the rigidity of the macro representation. None the less, this method of representation which effectively assumes a fixed turn-around-time for execution of process flow is an acceptable business process representation since each manufacturing area commits to an agreed turn-around-time and does not wish it to be undermined. The method for calculation of the resource implications was originally developed and deployed by Leachman *et al.* (1996) in the Berkeley Planning system designed for such production environments, and dubbed Dynamic Product Functions. The implication calculation process can be paraphrased in the following manner:

If I start production of one unit of product p on route r in time period t, assuming given product steps cycle times, and nominal step-to-step queueing delays and step yield, what would be the schedule of resources demands in periods $T \geq t$, and schedule of product output in periods $T \geq t$?

This information, referred to as resource_map and out_map, is then used to relate periodic start production decisions to resource constraints as well as output consequences.

In our implementation the highest level of representation is a sub-route in a given manufacturing stage, and we view each production activity as a *transformation*. Each transformation is a tuplet of the type (*parent_product*, *route*, *child_product*) which, implies that the parent_product via a given route creates the child_product. Since there are multiple facilities that can perform the same activities, and products produced in one stage may be moved to a choice of facilities for next stage of completion, we extend the transformation tuplet to include locality of each product: (*parent_product_area*, *parent_product*, *route*, *child_product_area*, *child_product*) conveying an inter-facility transfer when parent and child areas are different, including transshipment component of products from location to location. The transformation tuplet coupled with a time index (parent_product_area, parent_product, route, child_product_area, child-product, time) provides the necessary infrastructure information for key model decision variables. In our schema the tuple thus defined is referred to as the *connection* data, and at the data level the time index signifies the real date of the connection. By 'convention' we mean a strict *pull* interpretation of the connection data. Namely, the child product belonging to the child_area is created by pull of the parent product from the parent_product area, via the given route. As such the child product thus produced belongs to the child_product manufacturing area, and is inventoried there until pulled by a subsequent manufacturing area. The inventory balance equations are therefore produced for the child_product and the child_product manufacturing area. These inventory points are referred to as corporate inventory points (CIPs).

For a connection to be valid, at least one component of the pair area and product must be different between parent and child. Certain connections may be simple in the sense of the route involved – i.e. the route may simply be a shipment task for transfer of inventory from one location to another. It is clear that connection data between two products may not be singular and multiple alternatives (varying routes) may be used.

Figure 10.2 shows the connection between two consecutive stages. By convention, the product inventory point is associated with the area(s) where it is produced. In some instances exceptions occur, most notably when there is WIP in transit between two areas. In this case a new inventory point for the arriving parent product at the child product area

Figure 10.2 *Stage-to-stage connections*

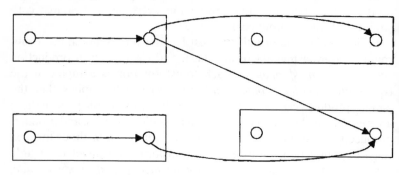

is needed, along with a new form of connection that is entirely in one area. While such instances occur commonly, they are not persistent throughout the planning horizon, and such cases occur only for a few initial time periods. For this reason, we chose to represent them as exceptions.

Given the basic interconnection data we can directly associate decision variables with each connection data representing start of production decision. The decision variable

S(parent_product_area, parent_product, route, child_product_area, child_product, time_period represents quantity of the production_start along the connection instance for a given planning time_period. Invocation of such a variable would signify production start of the child product in the child_product_area using as input parent_product from the parent_product_area in the specified time_period, causing reduction of the parent_product inventory in the given time_period, and increase of the child_product inventory in one or more future time_periods. The connection data thus provides a canonical form for the data-driven representation of production decision variables and inventory points. This is a nice and plausible form for the simple case of one-one transformations. However, with bining, and when a set of input products are required as input to create a subsequent product the basic connection structure becomes inadequate. With the aid of diagrams we now build-up the logical view of one-to-many (fission) and many-to-one (fusion) transformations with the introduction of *links*. Unlike connections which are route-based transformation, links are numerically-based associations between product pairs.

Figure 10.3 shows (a) a basic one-to-one transformation of product *i*

Figure 10.3   Product-to-product transformations

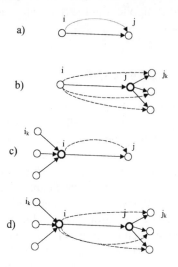

to *j* via route *r*; (b) shows a fission case where the child node *j* is a *logical* product linked to many real products j_k representing bins of *j*. Note that when bining occurs all the children are created automatically as siblings according to a statistical ratio representing the link attribute. (c) shows a many-to-one transformation where the child product *j* is created from a collection of parent product i_k that are linked to the logical product *i* with the link attribute representing the count of each component needed. (d) is the general case. The "many" set represents a set of products that are in "and" association. The logical products are shown as shaded nodes "linked" to another product. The logical products are part of the AMD's modern product nomenclature structure and are explicitly defined because of business needs. In case of fission logical nodes allow time-phased bining attributes without modification of product structure. On the fusion side, the collection of parent products is known as an assembly kit, and in practice alternative assembly kits may be used to create the same end product – e.g. a 32 meg memory module may be made of four 8-meg memory chips or eight 4-meg memory chips.

Using the extended interconnection notion of connections and links, the number of decision variables remains consistent with the number of connections, the inventory balance equations consistent with the number of real (not logical) products, and a singular data-driven method for variable and constraint generation becomes possible. As it will become evident, this is very important for large-scale modeling as well as an important component of the data validation process for large scale representations.

III. PRODUCTS AND DEMAND

In this environment the products' BOM consists of three distinct product types with varying requisite attributes. First are manufacturing products, products that are produced and managed in the supply chain. Second, materials which are purchased and consumed in the supply chain, and third the ordering products, which are products against which customer orders are taken. Each saleable manufacturing product has a link or an association with an ordering product. From a products' structure graph perspective, ordering products appear as pendant nodes with no products emanating from them. There is no requirement that a saleable product to be at the "end" of the supply chain. The manufacturing products to ordering products relationship is a many-to-many relationship, several manufacturing products may be used to define and fulfill an ordering product. A statement of demand is associated with ordering products, which have a selling price and a backlog penalty. The set of ordering parts may change significantly, based on the occurrence of custom orders that are created from manufacturing parts with different economic attributes.

Demand is stratified to consist of several demand classes or priorities, such as backlogs, order-book demands, inventory replenishment, or forecast demands. There is a strict observance of demand priorities – namely, no demand of a lower priority will be addressed before a demand of a higher priority no matter the economic consequences. Each demand class may have different objective measure. The process by which demand classes are addressed is iterative, where at each iteration a specific demand class is solved while the level of demand satisfaction from a previous iteration is forced in.

The fundamental cost factor in the planning model is the transformations cost, specifically the connection costs which are the comprehensive cost of creating a manufacturing product at a given stage. Transformation cost includes transportation and materials costs but not the cost of the parent product. Real manufacturing products, however, are given an inventory carrying penalty. All carrying and backlog costs are discounted and are used as incentives for just in time production and delivery of products.

IV. PLANNING HORIZON AND TIME-PHASED DATA

Given the characteristics of the production supply chain and business requirements the planning horizon is over a year long. The number of planning periods has a direct impact on the model size; large planning periods, such as weekly or monthly periods, are too approximate, while shorter-length planning periods create too many periods. The most reasonable approach is a planning horizon that starts with small periods and graduates into larger periods in a monotone nondecreasing fashion. This permits a greater time resolution for decisions in the execution time frame, and approximate results for future periods when demand is mostly forecast. The actual implementation of the application allows user definition of planning periods. However, for official periodic planning exercises which are repeated every week a rolling horizon of a fixed format is used. This horizon starts with at most 2 weeks' worth of daily planning periods, followed by 6 months' worth of weekly buckets, a few monthly buckets and one quarterly bucket. AMD deploys its own manufacturing calendar of working days for all the manufacturing facilities at its disposal. All planning exercises and measurements are based on the AMD working weeks/days, and the AMD calendar provides a mapping between each calendar day with AMD calendar. Hence, for a given day, one can determine the association of the day with AMD calendar week, month or quarter. For each day the number of working hours is defined by the manufacturing facility calendar. In each regenerative periodic planning run the composition of the planning horizon is modified to match the planning periods with AMD calendar weeks, months, and quarters.

The majority of the data used for planning is time-phased, all parameters change as a function of time, yields, cycle times, and even product structure changes with time. Such changes from a database perspective are managed by assignment of an effective date to the data element. This time is also represented in the interconnection data, however, with some interim data calculations certain procedural assumptions are made and deployed. Consider the data-generation process associated with implications of a given production start decision for a given connection instance. A connection defines a product pair and a route for transformation of the parent product to the child production. For the connection to be valid at a particular instance both the product structure and the route must be valid at the given instance. A route is defined as a sequence of manufacturing operations or steps. Each step has a cycle time, yield, and associations with one or more resources.

Figure 10.4   *Mapping of start interval to out interval*

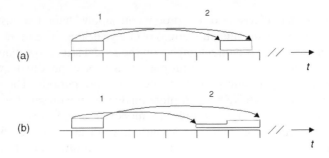

There is also an assumed inter-step delay or queueing time. The overall route cycle time and yield are calculated, based on the step information. With step information being time-phased, the route-level information become dependent on when in time the step is executed, which in turn becomes dependent on when the route was initiated. Prediction of a route completion time therefore becomes conditional on the route initiation time, and for each start decision for each start time period we have to calculate the corresponding route completion time schedule based on the applicable manufacturing area calendar and planning horizon. Figure 10.4 depicts the starts-to-outs mapping based on a uniform start footprint filling the start bucket. With varying size buckets (periods) various out maps are possible where the out footprint may be fully contained in a period, or span several periods. We also make the assumption that output occurring in a given period is uniformly distributed over the out period. The implication of the out footprint is shown in Figure 10.3 (b). While this assumption may be physically incorrect, a corrective measure in case of large time periods is deployed to insure that output availability in a larger bucket can be supportive of uniform availability in the period. A large period potentially has arrivals from starts of multiple earlier and possibly smaller periods. By analyzing the start to out map of adjacent time periods that produce output in a common larger period we provide data to be used in constraints relating the start quantities in such periods such that an earlier period is forced to produce more than its next adjacent period. In this manner the out period is loaded in a monotonically non-decreasing fashion.

Given that the time-phased data, mapping of the start footprint could potentially lead to certain anomalies when the data is changing during the days within a period. For example, the starts occurring at the end of the period may finish earlier than the starts at the beginning of the period causing the lines 1 and 2 in Figure 10.4 to cross. Several methods for calculation of the out map from highly accurate to less accurate may be deployed. Since the data may change at best at a daily resolution, one can calculate a daily out map, and for any period of any size create an out map based on aggregation of data associated with the days of the bucket. At the other extreme one may use the data instance associated with the beginning of the period for all of the period. While a method by averaging the data at the beginning and the end of the period may work for the out mapping, as will be shown in the next section it is not applicable with resources implication mapping. The code developed for calculation of the out map is very efficient and has linear order behavior applicable to either extreme. The choice is based on the nature of the data and the performance requirements of the overall application. Since planning periods are long and we are doing event-based planning, we qualify the notion of the start production quantity in a given planning period to a uniform start rate during the planning period.

V. RESOURCES MODELING AND THE RESOURCE MAP

Individual manufacturing resources are defined by each facility, and each resource has a net availability in terms of number of operating hours per day. Resources are defined by type, and each type has a resource count. Resources capable of similar tasks are not grouped into a singular type due to performance differences. Only resources that have homogeneous behavior define a type. A given manufacturing step (operation) may require several resource types simultaneously – e.g. a test operation may require a tester, two tester handlers, and one opera- tor. When the operation is engaged all the resources are simultaneously captured by the operation. Such a collection of resources is defined as a work-group. Manufacturing engineering determines the set of work- groups in a facility, the valid step to work-group associations, and unit per hour of the step by work-group. Because various generation of tools and resources may coexist, scheduled to arrive, or to be phased out, each step may have several choices of work-groups across time. Not all work-groups are modeled for capacity representations, only those with critical resources. If a resource type is declared critical, then all work-groups that contain the specific resource type are marked as critical. A manufacturing step requires resource implication representation if all work-group alternatives associated with the step are marked. Having an alternative work-group creates an additional level of complexity to the model, requiring additional decision variables for work-group choices. Typically, the percentage of the activities that have alternative critical work-groups is very small, but they are quite important in capacity representation in certain facilities.

The resource_map is the schedule of the resource requirements of a start decision. In other words, if we initiate a unit of start along a connection in period t, what is the schedule of resources demand for all subsequent periods based on the route attributes and unit production per hour (UPH) factors? Similar to the out_map, for each effective connection instance the set of resource demands is calculated. If a step has alternative work-groups the resources demand coefficients are calculated. If a step has alternative work-groups the resources demand coefficients are calculated for each alternative. If several operations require the same resource in the same time period then their loading demands are combined. The form of the resource map is (parent_product_area, parent_product, route, child_product_area, child_product, start_time_period, step, work_group, resource, affected_time_period, load_per_unit_of_start). However, when no alternative work-groups are involved, step and work-group indices are removed. This separation is done for model-size efficiency. Materials' dependencies are treated separately from manufacturing products' dependency in the supply chain. The modeling approach for representation of materials is done via aggregate limit constraints that limit total cumulative production to the total aggregate material availability in each time period. One assumption that is used is that all material must be ready at the time of the production start. For most cases, the instances of the materials constraints are limited to the procurement lead time of the product. It is assumed that the material can be purchased for production beyond the material acquisition lead time.

VI. WORK-IN-PROCESS (WIP) REPRESENTATION

One of the major issues in modeling of such systems is the representation of WIP. We distinguish between manufacturing product inventories which are defined products at the end of a production stage and

WIP – that is, a manufacturing product that is on a particular step of a route within a stage. Unless there is nomenclature to represent the partially complete manufacturing entities, the step identifier is the only means for determination of the extent of completion of a manufacturing product. One issue in the representation of WIP is the degree of freedom in reassessing the WIP movement priorities. In an extreme case the WIP may be replanned (reprioritized) for its residual tasks just like any new starts, while in another extreme case the priorities are predefined and inherited. For modeling of WIP we present two alternatives. In alternative 1 we treat WIP in any production stage with higher priority than any new starts, and hence a separate treatment and function for WIP management is used. In alternative 2 we provide an integrated model for WIP planning and new starts planning, leading to a more flexible yet larger model. Under alternative 1 similar to Leachman *et al.*'s (1996) approach, WIP is 'flushed' to its completion stage and represented as projected inter-stage product inventory to the planning model. In this manner, the model makes decisions regarding WIP in terms of inter-stage movements. As part of the flush process the capacity consumption by WIP is subtracted from the net available capacity for planning of new starts, and consequently, WIP is given priority over new starts. This approach tends to ignore the interaction of WIP with real starts, especially when the WIP involves re-entrant flows to the same manufacturing resources. Typical biasing approaches may be used to cause partial correction based on anticipated new starts. The role of the flush process is to predict the schedule of the completion of WIP. Solution methods for the flush process observed are simple deterministic rules; the use of simulation models; or math models. In our approach, we use a linear programming (LP) model. The WIP output schedule is optimized based on the stage production demand. The stage production demands are determined based on output prediction of the planning model from a previous run. Clearly, the inter-stage demand may not match the actual WIP; this mismatch may be caused by discrepancy between planned starts and WIP yields observed. The LP model pulls WIP according to demand, and pushes the extra WIP at the fastest rate possible. This model is described later. The flush LP model, contrary to the planning model, is a fine-grain model with better representation of time periods, and unique to the stage under consideration. It is a large-scale capacitated multi-commodity network flow model, because of the number of time periods used. One unfortunate aspect of an LP-based flush model is uncertainty in number of time periods required to complete the flush. A generous number of time periods makes the model unnecessarily large: the model grows in number of rows and columns by the number of time periods used. To avoid this problem, we use time-horizon segmentation, where if at the end of the run there is unfinished inventory of WIP the process is repeated with the residual WIP status as input. Although a sub-optimal process, results are adequate while the emphasis is on accuracy of the near-term predictions.

Note that the flush process is done independently for each facility in an asynchronous format. A WIP flush module, as a stand-alone tool, is a useful utility for each manufacturing area as a predictive tool.

In alternative 2, the WIP and new starts are managed under one treatment and model. To do so the definition of start variables is extended to include a step index – i.e. the start of a production may occur at *step_0* (new starts), or any subsequent step where there is WIP. The inventory balance equations are also extended to support inventory balance at *step_l* for $l < n$. However, the nature of balance equation at *step_l* is restricted to outflows only. This level of modeling does not change the model format; it increases the model complexity and size to the extent of WIP. The majority of the burden for this implementation occurs at data generation for resource mapping and start-to-outs mapping based on partial routes. The amount of data associated with these maps has been a limiting factor for implementation of this alternative. However, our implementation of the mapping module readily supports the required functionality without a significant module performance impact.

VII. DATA SUPPLY

The most complex component of the planning application development is large-scale data supply and data maintenance. In our instance, the planning system database is not the system of record for majority of the data needed. Other corporate data repositories are the primary data source systems. Examples of such systems are product structure data systems, demand and inventory management systems, financial data systems, and shop floor data systems. In some instances, these repositories provide database-to-database access and in other instances file transfer methods are used for data communication. The outputs of the planning system results are communicated to the planning database from which user reports are created. The planning database system receives data from all such systems and maintains updated images for planning needs. As such, data maintenance for data completeness and validity is a challenge: production planning with incomplete data is impossible. One approach is to load the planning model with whatever data available, scrub the data for the valid complete components, and then proceed with model generation and solve processes. Typically such approaches create a great deal of waste and increase the planning cycle time. Incomplete product structures may not cause infeasibilities, they simply do not result in planned production. Without knowledge of such errors it is not clear whether a product was not planned for capacity or economic reason vs data discontinuity. Two fundamental rules in data acquisition and maintenance processes are followed in our implementation. First, only valid, useful, and fully 'connected' data is maintained in the planning database. By 'connected' we mean that all necessary information to fully plan a product from start to end is available – for example, you may not have a manufacturing product for which a route is not defined. Second, all data other than status data that is highly volatile is communicated on a net-change basis, and validated upon arrival. Therefore, if the state of data is kept valid and only valid net changes are applied the resulting state will be valid. This process requires a more elaborate data communication and load process, however it pays off in the long run. This approach has several advantages. First, the data communication volume is reduced, which is important with offshore data communication in a global manufacturing setting. Net-change communication minimizes the amount of data validation work. Only useful data is kept, hence the database table sizes is at the minimum required. The net-change process allows new additions, deletions, and modifications of existing data. Key to successful validation is treatment of product structure as core data and treatment of the product structure as a connected graph with a dummy singular root, as shown in Figure 10.1. In this structure only the root node does not have a parent, only ordering products do not have children. Manufacturing products are supposed to have children eventually leading to an ordering product. Otherwise, they represent valid but incomplete structures. By node-type casting – i.e. root product, manufacturing products, and ordering products – the node in-degree and out-degree data can be used for maintaining a connected structure: an arc may not be removed if it makes the in-degree of any manufacturing product zero. Clearly, when loading product structure for a complete chain the order of data is important; unfortunately, this method of supply cannot be guaranteed. For case of product structure data, for a given net-change data set, first, record additions are processed repeatedly. If on a given pass a new node or an arc cannot be added, the process stops. For the process of data extraction to build the planning model, essentially a backward search is used to pull the required product structures. The search starts with products in demand and finds the appropriate parent product. The process is repeated recursively until the root node is reached. In this manner, only the necessary and sufficient element are identified and pulled.

VIII. MODEL MATRIX GENERATION

Model generation, particularly for the initial prototyping activities where the functional features of the optimization are being debated, can be an expensive effort. Modern algebraic modeling tools provide significant leverage in rapid prototyping and evaluation of the proposed model features. After examination of a few available packages we decided on IBM's Easy Modeler (EM) (1994) modeling system. This is an algebraic modeling tool with a graphical interface for model specification and data input. EM operates on model-specific files that may be produced through the EM interface. It supports linear, integer, and quadratic modeling. The EM modeling system creates C code as a main or sub-routine that may be embedded in the user application. Users may create applications using the EM library of subroutines. EM provides driver mechanisms for operation with solver systems such as the IBM Optimization Subroutine Library (OSL) and the IBM Mathematical Programming System/Extended (MPSX). Otherwise, it creates an MPS format file to be used by other solver systems. To use EM for large-scale very sparse matrix generation, certain modifications to the tool at our request were made to the IBM EM development team and delivered. First, there was a need for an alternative other than a file-based data feed. This was accomplished by provision of user functions (hooks) for direct input of data emanating from the database. Second, modification of the looping constructs for row and column generation was needed to allow for data table-driven generation. The typical looping clause of the type

FOR index i; FOR index j; FOR data instance relation $d(i,j)$ = true

required $O(i)O(j)$ iterations with checks for data instance $d(i,j)$. This is an expensive operation when the order of the indices i and j are large and wasteful when data instances $d(i,j)$ are very sparse. In our case, the main decision variable of the model, production start decisions has seven indices ranging in cardinality from $O(n^2)$ to $O(n^4)$, causing significant performance problems in generating large models. The modification requested and implemented provides for optional data_table-driven looping clauses such as *FOR d(i,j)*. In this manner, the number of iterations is limited to the number of records in the table, and the indices are supplied by records.

IX. HARDWARE/SOFTWARE PLATFORM

The choice of the hardware and the solver system for the application was based on a competitive benchmark study. For this purpose a large-scale model using randomly generated data based on a representative supply chain structure was produced by AMD using EM. The resulting matrix had nearly 550 000 columns and nearly 500 000 rows. Only the MPS file with no additional information was supplied to the vendors representing HW/Solver platforms. Based on the results delivered we chose IBM SP2 with Parallel OSL as our solution system. The database system is an Oracle DB server on a SUN system. All components of the application related to data acquisition and modeling support are written in C^{++}.

X. SOLUTION STRATEGY, DECOMPOSITION

Mathematical decomposition methods are typically used for solution of structured large-scale problems. One such method is the Dantzig–Wolf decomposition for block angular models with coupling rows. This observation provides a natural decomposition method that is well aligned with AMD's planning problem. Essentially, AMD is a collection of separate business entities called product lines, each with its own network of inter-connected products. The only interaction among product lines is sharing of the manufacturing capacity and material resources. In fact, if each product line owned its own share of production capacity a fully independent planning sub-system, one for each product line, could be exercised. Decomposition along product lines is a plausible level of decomposition, especially if as a by-product a capacity wedge may be obtained to be dedicated to a product line for independent exercises. A finer-grain decomposition is also possible by examination of each product tree. However, given that there are roughly 10 product lines and the business process implies a product line-based decomposition, this approach is used for our solution strategy. The implementation used is similar to the work of Ho and Etienne (1981).

XI. MODEL DATA REQUIREMENTS

The model data requirements are described below. We started by definition of sets and their index representation(s) and defined the data per the indices described. Certain sets are ordinal and are specified as such. Note that most data sets are super sparse, and only nonzero elements are supplied. The biggest effort of this application development has been data generation for the model using data stored in the data base. The rule followed in the design is 'minimality and sufficiency'. While it is possible that null data will have no effect in solution results, and will be cleaned by any solver presolve routine, it certainly will affect the model build time. The data-generation process is quite procedural to insure that the minimality rule is achieved. For example, the first step is processing of the demand data, which in turn is used to define proper the sub-set of the parenting manufacturing products in a recursive form. Along this process proper connection and link data are identified and extracted. Once the necessary core data is obtained, it is then expanded from time-phased effective data to periodic data based on the planning horizon, and manufacturing area calendars.

11.1. Sets

s sites   The list of manufacturing sites in AMD or beyond.

q,qq places   The list of manufacturing facilities in AMD and the vendor facilities. Data sets defined later provide associations between manufacturing places and sites. A site may contain several manufacturing facilities. Separation of manufacturing facilities within a site allows specification of plant specific features. While plants may not share equipment, they may share resources, such as materials, and operate under a common site calendar.

c classes   The list of various priority demand classes. This is an ordinal set. The classes are used for prioritization of demand, hence each demand data entry will have a class index. Demands of a higher class are processed before those of a lower class regardless of the economic attributes involved.

o objectives   The set of objective function components. This index is used to develop objective function components that are then used for various priority classes. For example, high priority demands are measured on the basis of the backlog penalties only.

t,tt,w time_periods   The ordinal list of time periods used for representation of the planning periods.

pr,pa,ch,gp,bp products   The list of manufacturing products in various stages of production. This list includes real and logical products.

opn OPNs   The list of ordering part numbers. Note that we have chosen to distinguish between manufacturing part numbers and ordering part numbers. It is perfectly legal for a name to appear in both, an ordering part number may be satisfied by many manufacturing part numbers.

f flows   The list of manufacturing methods, recipes that are used to transform products from one form to another. Flows are inclusive of shipment activities that permit transfer of products from site to site.

pl product_lines   The list of product lines in AMD. The index product lines allow separation of all products to various product-line groups. While grouping of the products may be done in a general manner via the grouping construct described above, due to pervasive use of this index and for efficiency purposes in model generation it is treated as a separate index by itself.

m materials   The list of consumable material resources.

r cap_resources   The list of critical bottleneck resources across all facilities under AMD control.

wg work_groups   The list of work-groups that contain bottleneck resources.

j steps_with_alts   The list of flow steps that have more than one workgroup attached to them. This is needed for modeling of alternative resources when certain manufacturing activities may be done in more than one way.

11.2. Data

The data section of the model provides data over the set member instances defined in the 'set' section:

place_site(q,s)   indicates association of a manufacturing facility with a site. Such association data sets are indicator data sets. To indicate a valid association a numerical value of '1' is used to pair a set of index instances. As a general rule only the nonzero instances are specified.

kit_link(pl,q,pa,ch)   defines the BOM components of an assembled product. Note that in this case the child product **ch** is a logical product, and represents a collection of products required to enable creation of a real singleton product.

opn_link(pl,q,pr,opn)   provides association between a manufacturing part number and an ordering part number.

cnxn(pl,q,pa,qq,ch,f,t)   provides product-structure information. The index product line is used for association of product with a product line. The pair **(q,pa)** defines the parent product and its location. Similarly the pair **(qq,ch)** defines the resulting child product and its location. '**f**' is the index of the flow or method of transforming the parent product to the child product with location considerations, and **t** is the time period when the connection is effective. Note that in connection data at least either product names or location names must change between the pairs. Mapping a product to itself at the same place has the effect of a self-loop. If the parent and child product names are the same but the locations are changing, then the anticipated effect is movement of a product from one place to another. The connection data set also covers the instances where the parent product is already at the child location. Such situations may also be mitigated by in-transit WIP. The data set **cnxn** is the most critical data set communicated to the model and is the key driver in creation of decision variables in the model. It is essential that only valid and complete associations are communicated. If products can be transformed in more than one way then each instance requires a proper flow index.

cost(pl,q,pa,qq,ch,f,t)   provides the cost associated with a manufacturing transformation activity. Hence, each activity in the supply chain should have a realistic cost number to allow differentiation in choice of product production or choice of the best place and method of producing it.

res_map(pl,q,pa,qq,ch,f,t,r,tt)   is a calculated data set deploying dynamic production functions. This data set conveys the resource impacts of start of a production activity under a given flow for transformation to another product. The time period **t** indicates the start period and the period **tt** indicates when resource **r** is impacted. The value of the data indicates the rate of capacity demand per unit of start of the transformation. This is a fairly large data set, since it is created for every valid connection data set and for every valid start period in the planning horizon.

alt_map(pl,q,pa,qq,ch,f,tt,j,wg,r,t)   is similar to resource mapping above except for the cases where alternative work-groups exist. Note that for steps that have alternative work-groups there is no mapping data in the **res_map** data set.

out_map(pl,q,gp,qq,ch,f,t,tt)   provides the relationship between the start of a transformation in a period **t** and the time period **tt** when the fraction of the start is completed. For any given valid start period there may be many out periods with varying values. The starts to out mapping calculator determines this data based on route cycle time and yield.

capacity(q,r,t)   provides the net general capacity of resource **r** at a particular place in a given time period. Note that the capacity amount is a function of time and size of the time period specified. The notion 'general capacity' signifies that this capacity bucket has not be pre-allocated and is open for consumption for any applicable transformation. This capacity is a netted value on the basis of the capacity consumption by WIP and any other reservations that may have been made against this capacity. Note that under the sparse data assumption convention, an unspecified capacity value is not the same as zero capacity value. Zero capacity cases must be explicitly defined. Otherwise, lack of specification means infinite capacity.

material_supply(s,m,t)   specifies the cumulative supply of material **m** up to time period **t** from beginning of the planning horizon. This is the unallocated availability of materials. A null specification means infinite supply. Typically, this data set is specified for time periods up to the procurement lead time of the material after which, it is unspecified, signifying infinite availability of the material. Note that this data is associated with the site, not the plant.

demand(pl,c,opn,t)   provides prioritized demand accumulated by class for **opn** products.

class_obj_assoc(c,o)   indicates association of a demand class and an objective function choice.

opn_valu(pl,opn,t)   is the unit value of each **opn** by time period.

prd_wip(pl,q,pr,t)   provides the projected WIP arrival date and quantity to the inventory point. This data is supplied as the result of the wip-flush process and covers in-transit items.

opn_inv(pl,opn,t)   supplies finished **opn** inventory that will be on hand at time period **t**.

matl_ascn(pl,q,pr,m)   defines the materials' requirements of the manufacturing products. The value signifies the number of pieces of material **m** required per unit of product **pr**.

alt_flow_steps(f,j)   is an indicator data set identifying steps on a route that have alternative work-groups.

step_workgroup(f,t,j,wg)   is an indicator data set specifying the time schedule of the step and work-group association.

huge   a very large number.

opn_volume(pl,opn)   represents total **opn** demand.

opn_active(pl,opn)   is an indicator data set for **opn** products in demand.

opn_bac_cost(pl,opn,t)   discounted backlog costs for **opn**s.

opn_inv_cost(pl,opn,t)   discounted **opn** carrying cost.

prd_place_ascn(pl,q,pr)   indicates valid locations where a product may exist. This data set is sub-set of the information contained in the **cnxn** data set. However, it is confined to set of real products, excludes logical products and is used for determination of inventory points.

cip(pl,q,pa)   is the set of inventory points.

prd_inv_cost(pl,qq,ch,t)   the discounted product inventory carrying costs.

back_bound(pl,c,opn,t)   assuming cumulative demand by class, the backlog control is defined per demand change by class using the current inventory and the previous class backlog solution.

bin_link(pl,q,pa,ch)   the data set **bin_link** defines the bin splits of the product with its associated bins. Note that in this case the parent product **pa** is a logical product, since it does not have physical reality, and **ch** is a real bin product. If a product under different testing methods can result in different bins or bin splits, then it is essential that a unique logical parent product **pa** be defined for each case. For products that have a unique bin, it is not required to have a bin link, although having one does not cause errors. Note this data set is produced from the **bin_split** data set **bin_split[pl,q,pa,ch,t]**, which provides the same information, however with the actual split ratio and in a time-phased fashion. Since the bin split data is folded into the out maps, it is not fed to the model directly.

11.3. Variables

The following is the list of the decision variables used in the model:
S(pl,q,pa,qq,ch,f,t)   production start variables signifying the decision

to start production of a child_product ending at location **qq** where (**qq,ch**) is a valid product inventory point using a valid parent product from location **q** where (**q,pa**) is a CIP, using route **f** in period **t**. Note that this variable matches the **cnxn** information of the product pair.

A(pl,q,pa,qq,ch,f,t,j,wg)   a decision variable for handling of the alternative work-group invocations.

I(pl,q,pr,t)   indicates the inventory level of a product at an inventory point for a given period. Only real products are modeled for inventory.

D(pl,q,pa,opn,t)   are distribution variables indicating allocation volume of a product to a valid **opn** based on a valid **opn_link**.

BL(pl,opn,t)   the variable backlog level of an opn with respect to demand.

IN(pl,opn,t)   the variable inventory level of an **opn**.

The following infeasibility variables (slacks) are needed in case of lower bounds that may be specified. Lower bounds may be infeasible. Hence a slack variable is used to insure feasibility. However, these slacks are heavily penalized in the objective function, so that they are avoided unless the bounds are infeasible.

O(pl,o)   a dummy variable used to capture the objective function value for a given solution. Instances of **o** are service and economic.

11.4. Formulation

Only some of the key components of the model are presented here. The actual model supports a greater deal of functionality, in particular with respect to inventories and other controls which would requires describing a larger set of data definitions.

11.5. Materials limitation constraints

For each instance of the material limit specified, all production starts of products that require the given material are bounded by the material availability. Since materials can be transferred from one period to the other, the logic requires that all production starts prior to a given period are limited to the cumulative supply at the time period. Note that the clause structure of the start variables is based on product and material association and manufacturing place and site association, as well as the proper **cnxn** data that creates a valid start variable. The material is attached to child-product and is assumed to be available at start of production regardless of the flow time.

$$Gen\_mat\_lmt(\forall\ material\_supply(s,m,t))$$
$$\sum_{pl,p,pa,qq,ch,f \in A} matl\_ascn(pl,qq,cm,m)S(pl,q,pa,qq,ch,f,tt)$$
$$\leq material\_supply(s,m,t)\ (for\ A = matl\_ascn(pl,qq,cm,m),$$
$$place\_site(s,qq),\ cnxn(pl,q,pa,qq,ch,f,tt),(tt \leq t))$$

(10.1)

11.6. Capacity constraint for general resources

The constraint is created for every capacity data instance supplied. Because a production start in period **t** may imply capacity demand in period **tt**, we must look for all production starts in periods **tt** that may demand capacity in period **t**. The proper sub-set of the production starts possibilities is identifiable from the **res_map** data set, hence it is used as a clause in selection of the qualified start variables that are in need of resource **r** capacity in period **t**. Note that an unspecified capacity limit would not result in a constraint generation which means unlimited capacity.

Let us introduce the following sets:

$$w = \{pl,q,pa,ch,f,tt\}$$

and

$$z = \{pl,q,pa,ch,f,j,wg,tt\}$$

then the capacity constraint becomes:

$$Gen\_resource(\forall\ capacity(qq,r,t))$$ (10.2)
$$\sum_{w \in A} res\_map(w,qq,r,t)S(w,qq)\ + \sum_{z \in B} alt\_map(z,qq,r,t)$$
$$A(z,qq) \leq capacity(qq,r,t)$$
$$for A = cnxn(w,qq)$$
$$for B = cnxn(w,qq)$$

11.7. Alternate work-group constraint

This constraint insures that start decision is broken into various alternative loadings

$$Alternate\_join(\forall \ caxn(pl,q,pa,qq,ch,f,t) \cap steps\_with\_alt(f,j))$$

$$\sum_{wg \in A} A(pl,q,pa,qq,ch,f,t,j,wg) = S(pl,q,pa,qq,ch,f,t) \qquad (10.3)$$

$$(forA = step\_workgroup(f,t,j,wg))$$

XII. OUTPUT MAPPING

Since the main decision variables of the model are production starts, to determine the outs we use nonbinding constraints to measure the outs as a function of starts. For every valid CIP which is defined by the data set **prd_place_ascn** and every time period **t**, we collect all production starts at periods **tt** prior to **t** such that their outs occur in period **t**. The **out_map** data set provides the proper sub-set of the variables that qualify for each instance. To the planned out the inventory due to WIP is added to report one total number. Since all production start variables are nonnegative, the lower bound of zero is always feasible. The first component covers increases from upstream activities connected to the child-product while the second one covers increases from upstream activities linked to this product

$$Outs(\forall \ cip(pl,ch,q),t)$$

$$\sum_{qq,pa,f,tt \in A} out\_map(pl,qq,pa,q,ch,f,tt,t)S(pl,qq,pa,q,ch,f,tt) \qquad (10.4)$$

$$+(forA = cnxn(pl,qq,pa,ch,f,tt),$$

$$\sum_{qq,gp,pa,f,tt \in B} out\_map(pl,qq,gp,q,ch,f,tt,t)S(pl,qq,gp,q,pa,f,tt)$$

$$+wip(pl,q,ch,t) \geq 0$$

$$(forB = bin\_link(pl,qq,pa,ch),cnxn(pl,qq,gp,q,pa,f,tt))$$

XIII. PRODUCTION INVENTORY BALANCE EQUATIONS

These production inventory balance equations are created for real manufacturing products at valid CIPs. The equation logic determines product inventory point as imbalance of input and output from the CIP. If input exceeds output from the CIP then inventory is created. These balance equations do not cover **opn**s, as they are treated separately. There are two cases, one for the initial time period where there is no inventory carryover from the previous time period, and one for the subsequent periods. The inventory increase can occur in two ways. First, if a product is descendent of a real product – i.e. there is direct connection to the product. Second, the product is created as result of a bining process, where there is link to the product, not a connection. Inventory decrease at a CIP occurs similarly by starts that take place from the CIP via a connection, or by other starts downstream that are connected to the inventory item via a link. In case of linked products the out maps are modified to associate relationships of real to real products. However, the distinction is made clear in use of the link data. Finally, inventory decrease can occur if the product is parent to an **opn**. For **t** = 0 the negative inventory time index is avoided

$$Inv\_balance(\forall \; cip(pl,ch,q),\, t)$$

$$I(pl,q,ch,t) = I(pl,q,ch,t-1) + new\_wip(pl,q,ch,t) +$$

$$\sum_{qq,pa,f,tt \in A} out\_map(pl,qq,pa,q,ch,f,tt,t)S(pl,qq,pa,q,ch,f,tt) +$$

$$(forA = cnxn(pl,qq,pa,ch,f,tt))$$

$$\sum_{qq,gp,pa,f,tt \in B} out\_map(pl,qq,gp,q,ch,f,tt,t)S(pl,qq,gp,q,pa,f,tt) +$$

$$(forB = bin\_link(pl,qq,pa,ch),cnxn(pl,qq,gp,q,pa,f,tt))$$

$$\sum_{opn} opn\_link(pl,q,ch,opn)D(pl,q,ch,opn,t) -$$

$$\sum_{qq,bp,f \in C} S(pl,q,ch,qq,bp,f,t) -$$

$$(forC = cnxn(pl,q,ch,qq,bp,f,t))$$

$$\sum_{pr,qq,bp,f \in D} kit\_link(pl,q,pr,ch)S(pl,q,pr,qq,bp,f,t)$$

$$(forD = cnxn(pl,q,pr,qq,bp,f,t)) \tag{10.5}$$

XIV. DEMAND SATISFACTION AND BACKLOG MEASURES FOR **OPNS**

These constraints are generated only for the **opn**s. Shortage of supply with respect to demand creates a backlog. The backlog from one period is carried to the next period similar to inventory

$Sales\_backlog(\forall\ opn\_active(pl,opn),t)$

$$BL(pl,opn,t) - BL(pl,opn,t-1) - IN(pl,opn,t) + opn\_inv(pl,opn,t)$$
$$+ \sum_{q,pr \in A} D(pl,q,pr,opn,t) = demand(pl,opn,t)$$
$$(forA = opn\_link(pl,q,pr,opn)) \tag{10.6}$$

$Production\_cntl(\forall\ opn\_active(pl,opn))$

$$\sum_{q,pr \in A} D(pl,q,pr,opn,t) = \sum_{\forall\ pl,opn,t} demand(pl,opn,t)$$
$$- \sum_{\forall\ pl,opn,t} opn\_inv(pl,opn,t)$$
$$(forA = opn\_link(pl,q,pr,opn)) \tag{10.7}$$

XV. OBJECTIVE FUNCTIONS – SERVICE COMPONENT

The measure of service is determined by the degree of backlog created with respect to demands that are in the 'service' priority class. For efficiency purposes, the constraint clause is limited to **opn** that are actively demanded

$$Objective(\forall\ pl;\ ``o" = `service')$$
$$O(pl,o) = \sum_{opn,t} opn\_bac\_cost(pl,opn,t)BL(pl,opn,t) \tag{10.8}$$

XVI. OBJECTIVE FUNCTION – ECONOMIC COMPONENT

The economic measure of the solution is determined in term of revenue-generating activities measured by distribution variables which allocate supply to **opn** demands and cost-incurring activities such as transformation costs, cost of carrying manufacturing products and carrying inventory of **opn**s. Also for infeasibility slacks an artificially high penalty is used to discourage their occurrence if possible

$$
\begin{aligned}
&\textit{Objective}(\forall \ pl; \ "o" = 'economic') \\
&O(pl,o) = \sum_{pr,opn,t \in A} opn\_valu(pl,opn,t)D(pl,q,pr,opn,t) - \\
&\qquad\qquad \sum_{opn,t} opn\_inv\_cost(pl,opn,t)IN(pl,opn,t) - \\
&\qquad\qquad \sum_{pr,q,t \in B} prd\_inv\_cost(pl,pr,t) * I(pl,q,pr,t) - \\
&\qquad\qquad \sum_{pr,pa,qq,ch,f,t \in C} cost(pl,q,pa,qq,ch,f,t)S(pl,q,pa,qq, \\
&\qquad\qquad\qquad ch,f,t) \\
&\qquad (\textit{for } A = opn\_link(pl,q,pr,opn),opn\_active(pl,opn)) \\
&\qquad (\textit{for } B = cip(pl,pr,q)) \\
&\qquad (\textit{for } C = cnxn(pl,q,pa,qq,ch,f,t)) \qquad\qquad (10.9)
\end{aligned}
$$

XVII. LIMITS

The bound section of the model provides constraints on individual variables. The first set of bounds are on the backlog variables of the **opn** products. This bounding process is required for maintenance of the priority demands as new demand classes are processed. Initially at the first demand class the bound is set to a large number to render the bound ineffective. As new classes of demand are introduced the value of the bound is modified to the backlog level at the previous demand class plus the maximum backlog that can be caused by the new demand class. The clause used is all actively demanded **opn**s and time periods

$$
\begin{aligned}
&\forall opn\_active(pl, opn),t \qquad\qquad\qquad (10.10) \\
&BL(pl,opn,t) \leq huge
\end{aligned}
$$

Changing the lower limit of the economic variable to a negative value from default of zero, since the economic measure may become negative

$$\forall pl, \text{``}o\text{''} = \text{`economic'} \tag{10.11}$$

$$O(pl,o) \geq - huge$$

All other variables are assumed to be nonnegative.

XVIII. OBJECTIVE FUNCTION USED

Since a linear program can have only one objective function, the objective function used is the difference of the economic and service measure. Initially for service demand classes the economic component is turned off, and later with economic demand classes it is turned on, the objective being maximization of the net profit

$$\max \sum_{\forall pl, \text{``}o\text{''}=\text{`economic'}} O(pl,o) - \sum_{\forall pl, \text{``}o\text{''}=\text{`service'}} O(pl,o) \tag{10.12}$$

XIX. REVISIONS AND ITERATIONS

The model described above is set-up for revisions and resolve. The original model created in memory is initially set-up so that it may have the maximal set of rows and columns that may be needed to handle all subsequent models for multiple demand classes and other functional features that are invoked at various iteration classes. Demand classes is managed by change of the demand values. For this reason, the demand in each priority class is cumulative with respect to the prior demand classes. As each new demand class is tried the data representing the next class demand is edited along with the value of the backlog variables to protect demand satisfaction to the level achieved. If during the process the nature of demand requires a change in the objective function from service to economic, this edit is added too. Consequently, the initial model is designed on the basis of the demand requirements of the last demand class, and immediately edited to represent the first demand class. This is also why at the start of the model generation upper bound and lower bounds of huge and zero are used.

XX. WIP FLUSH MODEL

For the flush process the time periods are units which are multiple of hours, dividing a day into multiple time periods. For process modeling a route is segmented into blocks or sub-routes. Sub-routes are created by marking of the steps that hit critical bottleneck resources. For steps in between critical steps, we assume that work can proceed at a nominal rate. The collection of the steps in between critical steps is combined into one step. WIP that is initially on such steps is pre-processed and moved to the next critical step with a given lag.

20.1. Sets

t,s time_period   The ordinal index of time periods.
i,j product   The list of products.
l step   The list of route segments for all routes.
r resource   The list of critical resources.
g work-group   The list of work_groups.

20.2. Data

capacity (r,t)   The amount of critical resource capacity.
prd_wip(i,l,t)   WIP, by step and time.
prd_step_time(i,l)   Product step duration, in integer values.
prd_step_yield(i,l)   Product step yield.
prd_res_map(i,l,g,r)   Step resource requirement, by work-group.
prd_demand(j,t)   Product demand.
prd_cost(i)   Product cost.
prd_prd_ascn(i,j)   Product-to-product association. Typically a product is associated with itself – however, if there is bining, count explosion, or name change, the association is different. The attribute of the association defines the change is quantity.
final_step(i)   This supplies the index of the last step for a product.
c(j)   This signifies child products.
p(i)   This signifies parent products.
cum_d(j,t)   The cumulative demand for given demand instances and the last period.
prd_bac_cost(j)   The product unit backlog penalty derived from product cost.
prd_inv_cost(i,l)   A monotonically decreasing by index l product holding cost.

aa(i,l,g) Indicator data set for product step to work_group association.

20.3. Variables

X(i,l,t)   The quantity of WIP **i** processed at **t** at step l.
A(i,l,g,t)   Choice of the work_group.
I(l,t)   Inventory of **i** at step l at time **t**.
B(j,t)   Backlog of product **j** with respect to demand at time **t**.
O(i,t)   Units of product **i** completed by time **t**.
Y(j,t)   Allocation towards demand.
All variables are non-negative.

20.4. Constraints

Balance($\forall prd\_step\_time(i,l), t > 0$)

$$I(i,l,t) = I(i,l,t-1) + prd\_step\_yield(i,l-1)$$
$$X(i,l-1,t-s) - X(i,l,t) + prd\_wip(i,l,t)$$
$$(For\ s = prd\_step\_time(i,l-1), s <= t) \qquad (10.13)$$

Completions($\forall p(i), l = final\_step(i), t > 0$)

$$I(i,l,t) = I(i,l,t-1) + prd\_step\_yield(i,l-1)$$
$$X(i,l-1,t-s) - O(i,t)$$
$$(For\ s = prd\_step\_time(i,l-1), s <= t) \qquad (10.14)$$

Exits($\forall c(j), t$)
$$Y(j,t) = prd\_prd\_ascn(i,j)O(i,t) \qquad (10.15)$$
Backorders($\forall cum\_d(j,t)$)

$$B(j,t) + \sum_{s \le t} Y(j,s) \ge cum\_d(j,t); \qquad (10.16)$$

Resource_link($\forall prd\_step\_time(i,l), t$)

$$X(i,l,t) = \sum_{g \in aa(i,l,g)} A(i,l,g,t) \qquad (10.17)$$

Capacity($\forall r, t$)

$$\sum_{i,l,g} prd\_res\_map(i,l,g,r)A(i,l,g,t) \le capacity(r,t) \qquad (10.18)$$

minimize backlog_inventory_cost

$$\min \sum_{i,l,t,i \in p(i)} prd\_inv\_cost(i,l,)I(i,l,t) +$$

$$\sum_{j,t,j \in c(j)} prd\_bac\_cost(j)B(j,t) \qquad (10.19)$$

The solution method for the resulting model uses a product-level decomposition approach. In our current implementation we use OSL's Dantzig–Wolf decomposition to establish a good initial feasible solution with a portion of resources allocated to each product. Our experience has shown that for our problems the best strategy is to follow the initial solution with a steepest descent primal process. Typical problem sizes are in the order of 50 000 rows and columns, and highly sparse. Solution performance is highly dependent on the number of capacity constraints, and requires 10–20 minutes of CPU time and a single RS6000 120 MHz node. Clearly the model size is highly dependent on the number of time periods used. For very large-scale instances we intend to segment the model across time. The model is executed for a finite number of time periods; if at the end of the run there is unfinished WIP the problem is repeated with the WIP status at the end of the run as input. This is clearly a sub-optimal process, but knowing that accuracy of the results is more significant with initial time periods we do not see this approach as a great compromise. For the latest treatment of such large-scale problems see McBride and Mamer (1997).

XXI. COMPUTATIONAL RESULTS

At this time, limited computational results are available. For the WIP flush model the solution times using OSL's one pass of the decomposition routine with partitioned right-hand side has proven very valuable in solution time. For actual testing process a data-generation system has been developed that is capable of populating the planning database system with structured random data using specification parameters for size of the entities. A similar approach was used in generation of a large-scale model for the benchmarking process. The model required 6 CPU hours to complete on a six-node SP system using OSL's parallel interor method with a simplex cross-over. Experimenting with random data covering the Fab and Sort areas on models with 46 000 rows and 87 000 columns for one service and one economic iteration require less than 26 minutes of time for data fetch, model build, and solution time using the direct primal simplex method on a 70 MHz RS6000 single processor 390H system.

XXII. CONCLUSION

In this chapter we have describe major undertaking for development of an enterprise-wide planning system. In this effort management and supply of data has proven to be most challenging aspect of the project. Certain strategies such as design of a unified representation of the product structures have proven invaluable. Large-scale high-performance applications tend to stress every aspects of systems design, stressing sub-systems' performance beyond normal expectations. Examples of such cases have been enhancements to the EasyModeler system, as well as the need for support from IBM's OSL development team. As a strategic tool we plan to work on enhancements in particular with respect to performance.

Acknowledgment

We would like to thank Mr Randy Burdick, AMD's Vice President of ITM group for his continuous support of this project, Dr Stefano Gliozzi of IBM Rome for his support of EM, and Dr Robert Clark of IBM OSL development group.

References

Ho, J.K. and L. Etienne (1981) 'An Advanced Implementation of the Dantzig – Wolf Decomposition Algorithm For Linear Programming', *Mathematical Programming*, 20, 303–26 (Amsterdam: North-Holland).

IBM AIX EasyModeler/6000 (1994) *User Guide, Release 2.0*, SB13-5249, IBM SEMEA.

Leachman, R., R. Benson, C. Liu and D. Rear (1996) 'IMPRess: An Automated Production Planning and Delivery Quotation System at Harris Corporation – Semiconductor Sector', *Interfaces*, 26, 6–37.

McBride, R. and J. Mamer (1997) 'Solving Multicommodity Flow Problems with a Primal Embedded Network Simplex Algorithm', *INFORMS Journal of Computing*, 9, 2.

11 Cargo Analytical Integration in Space Engineering: A Three-dimensional Packing Model

Giorgio Fasano

I. INTRODUCTION

The *cargo analytical integration* of carriers and modules addresses two main activities: the *cargo accommodation* and the *engineering analysis*. The first focuses on the layout aspects, geometrical arrangement, and mass properties. The second deals with the compatibility analysis to ensure the feasibility of the cargo accommodation, the overall system functional and operational characteristics. Tight requirements of modularity, accessibility, operability, system static and dynamic balancing, logistics constraints, and functional and operational conditions lead to an efficient use of the available volume.

The International Space Station (ISS, see Brinkley *et al.*, 1997) is a major example of a cargo analytical integration. A fleet of transportation systems will upload the material requested on orbit. A high-level upload planning is then performed in order to define – for each launch and for each carrier – the mass types and quantities to be transported on orbit. A detailed analysis of each single cargo has to be made (in order to meet the upload plan). Individual items are selected from a group of possible candidates, depending on membership, availability, and operational priority, and their (analytical and physical) integration is performed.

The cargo accommodation relates to a three-dimensional object layout mainly limited by tight constraints. Traditionally, experienced designers performed this task 'by hand'. However the growing number of technical and financial constraints require efficient solutions.

Figure 11.1   An ATV

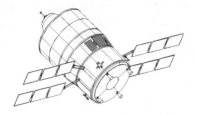

   The introduction of advanced methodologies and dedicated tools to support the cargo accommodation activity becomes a real necessity. This chapter introduces a specific three-dimensional packing problem defined as the allocation of a given set of parallelepipeds into a convex domain. The model is formulated in terms of mixed-integer programming (MIP) constraints. Even if no explicit target function is required, an ad hoc one is introduced to reduce the computational difficulties.

   Specific approaches including artificial intelligence or simulated annealing techniques (see Daughtrey *et al.*, 1991; Boeing, 1996) have been adopted to tackle different cargo accommodation issues.

II. THE CARGO ACCOMMODATION OF A SPACE CARRIER

The Automated Transfer Vehicle (ATV, see Amadieu, 1997) cargo accommodation in Figure 11.1 shows the European Space Agency (ESA) contribution to the mixed fleet provided by the international space agencies which will deliver to the ISS the resupply necessary to keep the Space Station operative and habitable. In this case, the cargo accommodation task consists of allocating 'small' items into containers (drawers, trays, soft packs), containers or 'large' items into racks and racks into (predefined) locations inside the ATV Cargo Carrier, taking into account mass and volume capacity limitations (at container, rack and cargo carrier level), specific positioning rules, and static and dynamic balancing requirements. Moreover, an efficient exploitation of the transportation capacity is necessary (Fasano and Provera, 1997).

   A straight formulation of the overall ATV cargo accommodation problem could have been given in terms of three-dimensional bin packing with extra conditions (Dyckhoff *et al.*, 1997). This approach, however, would have proved impracticable, because of the very large scale and highly *combinatorial* nature of the problem. A partition into a number of sub-problems has then been considered, following a bottom-up procedure (in combination with backward iterations when the convergence is not attained), and dedicated mathematical models, based on MIP, have been formulated (Nemhauser and Wolsey, 1988; Williams, 1993). A specific *three-dimensional packing model* concerning the allocation of a given set of parallelepipeds into a convex domain (with possible additional constraints) is reported in this work together with some practical results. The following paragraphs deal with the ATV pressurized cargo accommodation from the project led by Aerospatiale (Cannes, France) and Alenia Aerospazio.[1]

The ATV consists of a spacecraft, a pressurized and an unpressurized module. The resupply material uploaded to the ISS is generally denoted by 'pressurized', when transported by a cargo carrier that ensures a pressurized environment. Examples of pressurized cargo that can be transported inside the ATV Pressurized Module are:

- crew supplies (food, clothes, . . .)
- logistics items (for maintenance)
- science equipment
- oxygen for the extravehicular activity suit
- oxygen fuel
- the extravehicular suit.

For each mission, all items selected to be transported on-orbit can be 'small' and 'large'.

Containers can be defined as drawers, trays, and soft packs. A set of racks is given, as well as the set of possible *locations*. For each mission the following accommodations have to be performed:

- small item into a suitable container
- selected container and each large item into a rack
- rack into a location.

Table 11.1 reports the various ATV *accommodation levels*. It gives the correspondence between the Elements and the Accommodations.
Depending on the mission under study, a sub-set of *items* must be:

- incompatible with respect to others and cannot be accommodated into the same container or rack.
- accommodated in the same containers or rack.

Table 11.1 ATV accommodation levels

Elements	Accommodations
Small items	Drawers
Small items	Trays
Small items	Soft packs
Drawer	Racks
Trays	
Soft packs	
Large items	
Racks	(Pressurized module)
	Locations

Each container or rack has limits of volume and mass capacity and specific items must be:

- accommodated in specific containers or racks
- oriented in a predefined way
- with a fixed position
- no adjacent items
- 'piled'.

For each mission, moreover, each rack has to be *balanced statically* – that is, its center of gravity must lie in the neighborhood of an assigned point. The same requirement holds for the overall cargo system, in addition to the *dynamical balancing* – that is, the assignment of upper and lower bounds upon the inertia products, corresponding to a predefined reference frame.

Each item may be considered as a parallelepiped (or block of grouped parallelepipeds, in the case of particular large items) as well as each drawer, tray, soft pack (which is ideally enclosed in a box) or rack. It is assumed that all items and containers have to be positioned orthogonal with reference to the container or rack frame. Drawers and trays have predefined orientations, while any arbitrary (orthogonal) rotation is allowed for all items and soft packs, except in particular cases.

High flexibility in performing this kind of analysis and in evaluating the impacts of different options is necessary. During the development and utilization phases, moreover, quick planning and replanning activities are foreseen, with direct or indirect implications on the cargo analytical integration aspects. This is adequately supported by advanced methodologies, as well as appropriate tools.

The overall problem is tackled at different levels. One-dimensional bin packing models solve the assignment of small or large items to containers or racks, as well as the assignment of racks to locations. Once each small item has been assigned a container, or analogously each large item or container has been assigned a rack, a set of accommodation problems is set up.

A *three-dimensional packing model* represents these phases. If any accommodation problem at container or rack level proves to be unfeasible, some items or containers are rejected, following predefined selection criteria. Rejected elements are considered again for an alternative assignment to different containers or racks and a recursive procedure supports the overall cargo accommodation problem.

III. THREE-DIMENSIONAL PACKING MODEL

Three-dimensional bin packing problem consists of selecting a sub-set of parallelepipeds from a given collection to be accommodated orthogonal (with or without rotation) into a box in order to occupy the maximum volume. This problem is normally treated with heuristic or deterministic approaches and its usual formulation does not consider additional constraints upon the global center of gravity position.

Because of its combinatorial nature (Dyckhoff *et al.*, 1997), general-purpose MIP solvers are not able to attain a satisfactory solution. Difficulties arise when additional constraints are imposed, so that it is often necessary to implement dedicated algorithms.

The packing issue treated here is not formulated in terms of optimization model. The parallelepipeds have not to be selected from a given set. A mathematical model is expressed in terms of MIP constraints looking for a feasible solution. The model allow us to introduce some conditions (such as the *static balancing* of the entire system) to fix the position or predefine the orientation for some items. This model's extensions consist of adding constraints or fixing variables.

IV. MODEL FORMULATION

4.1. Indices and variables

Given the set of a parallelepipeds $P_1, P_2, P_i, \ldots P_n$, for $i = 1, 2, \ldots n$, let us denote by $L_{\alpha i}$, for $\alpha = 1, 2, 3$ its sides and by x_i, y_i, z_i its center of gravity coordinates with respect to a predefined orthonormal reference frame *XYZ*. A *convex domain D* is defined and the following positioning rules are stated.

The parallelepipeds should:

- have each side parallel to a frame axis (*orthogonal preorientation* constraints)
- be contained within *D* (*domain* constraints)
- not overlap (*nonintersection* constraints)

In order to formulate the positioning rules as MIP constraints we introduce the following *binary* variables:

$$\delta_{x\alpha i} \in \{0,1\}, \ \delta_{x\alpha i} = 1 \ if \ L_{\alpha i} \ is \ parallel \ to \ the \ X \ axis \ and$$
$$\delta_{x\alpha i} = 0 \ \text{otherwise} \quad \forall i, \ \forall \alpha$$
$$\delta_{y\alpha i} \in \{0,1\}, \ \delta_{y\alpha i} = 1 \ if \ L_{\alpha i} \ is \ parallel \ to \ the \ Y \ axis \ and$$
$$\delta_{x\alpha i} = 0 \ \text{otherwise} \quad \forall i, \ \forall \alpha$$
$$\delta_{z\alpha i} \in \{0,1\}, \ \delta_{z\alpha i} = 1 \ if \ L_{\alpha i} \ is \ parallel \ to \ the \ Z \ axis \ and$$
$$\delta_{x\alpha i} = 0 \ \text{otherwise} \quad \forall i, \ \forall \alpha$$

Used in nonintersection constraints, we introduce another set of *binary* variables: for any $i < j$, the set of *binary* variables:

$$\sigma^+_{xij}, \sigma^-_{xij}, \sigma^+_{yij}, \sigma^-_{yij}, \sigma^+_{zij}, \sigma^-_{zij} \quad \forall i, \forall j \ \text{and} \ i < j$$

where:

$$\sigma^+_{xij} = -1 \ \text{if the parallelepiped} \ i \ \text{comes before} \ j \ (x_i < x_J) \ \text{and}$$
$$= 0 \ \text{otherwise}$$
$$\sigma^-_{xij} = -1 \ \text{if the parallelepiped} \ i \ \text{comes after} \ j \ (x_i > x_J) \ \text{and}$$
$$= 0 \ \text{otherwise}$$

The same conditions hold for the other four variables.

4.2. Constraints

The *orthogonal preorientation* constraints are the following:

$$\delta_{x\alpha i} + \delta_{y\alpha i} + \delta_{z\alpha i} = 1 \quad \forall i, \forall \alpha \tag{11.1}$$

$$\sum_{\alpha=1}^{3} \delta_{x\alpha i} = 1 \qquad \forall i \qquad\qquad (11.2)$$

$$\sum_{\alpha=1}^{3} \delta_{y\alpha i} = 1 \qquad \forall i \qquad\qquad (11.3)$$

$$\sum_{\alpha=1}^{3} \delta_{z\alpha i} = 1 \qquad \forall i \qquad\qquad (11.4)$$

The *domain* constraints assume a very simple expression when D is a *parallelepiped* (see Figure 11.2). It is the situation supposed here, as the extension to the case of any other convex domain is quite straightforward. D is supposed to be oriented orthogonal with respect to the XYZ frame – that is, its sides D_x, D_y, D_z are parallel to the corresponding axes. A vertex of D is moreover supposed to be coincident with the frame origin O.

The simplified version of the domain constraints is reported below

$$0 \le x_i - 1/2 \sum_{\alpha=1}^{3} \delta_{x\alpha i} L_{\alpha i} \le x_i + 1/2 \sum_{\alpha=1}^{3} \delta_{x\alpha i} L_{\alpha i} \le D_x \quad \forall i \qquad (11.5)$$

$$0 \le y_i - 1/2 \sum_{\alpha=1}^{3} \delta_{y\alpha i} L_{\alpha i} \le y_i + 1/2 \sum_{\alpha=1}^{3} \delta_{y\alpha i} L_{\alpha i} \le D_y \quad \forall i \qquad (11.6)$$

$$0 \le z_i - 1/2 \sum_{\alpha=1}^{3} \delta_{z\alpha i} L_{\alpha i} \le z_i + 1/2 \sum_{\alpha=1}^{3} \delta_{z\alpha i} L_{\alpha i} \le D_z \quad \forall i \qquad (11.7)$$

Figure 11.2   Orthogonal three-dimensional packing into a parallelepiped

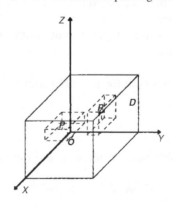

The *nonintersection* conditions (with the *orthogonal preorientation* rule) are equivalent to the *logical* condition below that must hold for any $i < j$

$$\left\{2|x_i - x_j| \geq \sum_{\alpha=1}^{3}(\delta_{x\alpha i}L_{\alpha i} + \delta_{x\alpha j}L_{\alpha j})\right\}$$
$$\vee \left\{|y_i - y_j| \geq \sum_{\alpha=1}^{3}(\delta_{y\alpha i}L_{\alpha i} + \delta_{y\alpha j}L_{\alpha j})\right\}$$
$$\vee \left\{|z_i - z_j| \geq \sum_{\alpha=1}^{3}(\delta_{zy\alpha i}L_{\alpha i} + \delta_{z\alpha j}L_{\alpha j})\right\} \tag{11.8}$$

This statement may be expressed in terms of MIP constraints. A *straight formulation* is attained by introducing the constants M^+_{xij}, M^-_{xij}, M^+_{yij}, M^-_{yij}, M^+_{zij}, M^-_{zij} (of appropriate values – e.g. M^+_{xij}, $M^-_{xij} \geq 2D_x$; M^+_{yij}, $M^-_{yij} \geq 2D_y$; M^+_{zij}, $M^-_{zij} \geq 2D_z$). The following expressions result:

$$x_i - x_j \geq \tfrac{1}{2}\sum_{\alpha=1}^{3}(\delta_{x\alpha i}L_{\alpha i} + \delta_{x\alpha j}L_{\alpha j}) - (1 - \sigma^+_{xij})M^+_{xij}$$
$$\forall i, \forall j \text{ and } i < j \tag{11.9}$$

$$-x_i + x_j \geq \tfrac{1}{2}\sum_{\alpha=1}^{3}(\delta_{x\alpha i}L_{\alpha i} + \delta_{x\alpha j}L_{\alpha j}) - (1 - \sigma^-_{xij})M^-_{xij}$$
$$\forall i, \forall j \text{ and } i < j \tag{11.10}$$

$$y_i - y_j \geq \tfrac{1}{2}\sum_{\alpha=1}^{3}(\delta_{y\alpha i}L_{\alpha i} + \delta_{y\alpha j}L_{\alpha j}) - (1 - \sigma^+_{yij})M^+_{yij}$$
$$\forall i, \forall j \text{ and } i < j \tag{11.11}$$

$$-y_i + y_j \geq \tfrac{1}{2}\sum_{\alpha=1}^{3}(\delta_{y\alpha i}L_{\alpha i} + \delta_{y\alpha j}L_{\alpha j}) - (1 - \sigma^-_{yij})M^-_{yij}$$
$$\forall i, \forall j \text{ and } i < j \tag{11.12}$$

$$z_i - z_j \geq \tfrac{1}{2}\sum_{\alpha=1}^{3}(\delta_{z\alpha i}L_{\alpha i} + \delta_{z\alpha j}L_{\alpha j}) - (1 - \sigma^+_{zij})M^+_{zij}$$
$$\forall i, \forall j \text{ and } i < j \tag{11.13}$$

$$-z_i + z_j \geq \frac{1}{2} \sum_{\alpha=1}^{3} (\delta_{z\alpha i} L_{\alpha i} + \delta_{z\alpha j} L_{\alpha j}) - (1 - \sigma_{zij}^-) M_{zij}^-$$

$\forall i, \forall j$ and $i < j$ \hfill (11.14)

$$\sigma_{xij}^+ + \sigma_{xij}^- + \sigma_{yij}^+ + \sigma_{yij}^- + \sigma_{zij}^+ + \sigma_{zij}^- = 1$$

$\forall i, \forall j$ and $i < j$ \hfill (11.15)

4.3. Model reformulation

The problem requires the conditions to be satisfied. The MIP iterations stop when the model is feasible.

In a consistent number of practical instances, however, it is hard to find an *integer solution*. The computational task becomes harder when the number of parallelepipeds involved grows and when the occupied volume approaches the available space.

A major difficulty is related to *nonintersection* constraints, which are also the biggest set of constraints in the model, being of the *fixed charge* type (Nemhauser and Wolsey, 1988; Williams, 1993).

In particular, the large value of the coefficients M_{xij}^+, M_{xij}^-, M_{yij}^+, M_{yij}^-, M_{zij}^+, M_{zij}^-, which acts as constraints bounds, reduces the computational performances.

Remarkable efforts in the area of MIP have been made on the issue of *bounds tightening* (Nemhauser and Wolsey, 1988; Williams, 1993). For *fixed charge* of large-scale problems, an efficient *preprocessing* technique has been implemented to reduce the fixed-charge bounds. The preprocessing compresses the region delimited by the fixed charge constraints in the linear programming (LP) relaxation (Nemhauser and Wolsey, 1988; Suhl, 1983; Williams, 1993). A specific approach has been adopted for the three-dimensional packing. It represents a very special case of fixed-charge problems, for which the above mentioned preprocessing techniques would not be directly applicable.

A reformulation of the nonintersection constraints is performed and an *ad hoc target function* introduced. The constant $M = \max\{2D_x, 2D_y, 2D_z\}$ is defined and, for any $i < j$, the variables $\lambda_{xij}^+, \lambda_{xij}^-, \lambda_{yij}^+, \lambda_{\overline{y}ij}^+, \lambda_{zij}^+, \lambda_{zij}^- \in [0,M]$ have been introduced.

The following constraints hold:

$$x_i - x_j \geq \frac{1}{2} \sum_{\alpha=1}^{3} (\delta_{x\alpha i} L_{\alpha i} + \delta_{x\alpha j} L_{\alpha j}) - M + \lambda_{xij}^+$$

$\forall i, \forall j$ and $i < j$ \hfill (11.16)

$$-x_i + x_j \geq \frac{1}{2} \sum_{\alpha=1}^{3} (\delta_{x\alpha i} L_{\alpha i} + \delta_{x\alpha j} L_{\alpha j}) - M + \lambda_{xij}^-$$

$$\forall i, \forall j \text{ and } i < j \tag{11.17}$$

$$y_i - y_j \geq \frac{1}{2} \sum_{\alpha=1}^{3} (\delta_{y\alpha i} L_{\alpha i} + \delta_{y\alpha j} L_{\alpha j}) - M + \lambda_{yij}^+$$

$$\forall i, \forall j \text{ and } i < j \tag{11.18}$$

$$-y_i + y_j \geq \frac{1}{2} \sum_{\alpha=1}^{3} (\delta_{y\alpha i} L_{\alpha i} + \delta_{y\alpha j} L_{\alpha j}) - M + \lambda_{yij}^-$$

$$\forall i, \forall j \text{ and } i < j \tag{11.19}$$

$$z_i - z_j \geq \frac{1}{2} \sum_{\alpha=1}^{3} (\delta_{z\alpha i} L_{\alpha i} + \delta_{z\alpha j} L_{\alpha j}) - M + \lambda_{zij}^+$$

$$\forall i, \forall j \text{ and } i < j \tag{11.20}$$

$$-z_i + z_j \geq \frac{1}{2} \sum_{\alpha=1}^{3} (\delta_{z\alpha i} L_{\alpha i} + \delta_{z\alpha j} L_{\alpha j}) - M + \lambda_{zij}^-$$

$$\forall i, \forall j \text{ and } i < j \tag{11.21}$$

Moreover, for any $i < j$ the following *logical* condition is stated

$$\{\lambda_{xij}^+ \geq M\} \vee \{\lambda_{xij}^- \geq M\} \vee \{\lambda_{yij}^+ \geq M\} \vee \{\lambda_{yij}^- \geq M\}$$
$$\vee \{\lambda_{zij}^+ \geq M\} \vee \{\lambda_{zij}^- \geq M\} \tag{11.22}$$

It is quite easy to prove that the constraints (11.9)–(11.14) and (11.15) are equivalent to conditions (11.16)–(11.21) and (11.21). Furthermore, *logical* condition (11.22) may be expressed in terms of MIP constraints.[2]

Let us introduce a set of binary variables $\rho_{xij}^+, \rho_{xij}^-, \rho_{yij}^+, \rho_{yij}^-, \rho_{zij}^+, \rho_{zij}^- \in \{0,1\}$ and the following constraints:

$$\lambda_{xij}^+ \geq \rho_{xij}^+ M \qquad\qquad \forall i, \forall j \text{ and } i < j \quad (12.23)$$

$$\lambda_{xij}^- \geq \rho_{xij}^- M \qquad\qquad \forall i, \forall j \text{ and } i < j \quad (12.24)$$

$$\lambda_{yij}^+ \geq \rho_{yij}^+ M \qquad\qquad \forall i, \forall j \text{ and } i < j \quad (12.25)$$

$$\lambda_{yij}^- \geq \rho_{yij}^- M \qquad\qquad \forall i, \forall j \text{ and } i < j \quad (12.26)$$

$$\lambda_{zij}^+ \geq \rho_{zij}^+ M \qquad\qquad \forall i, \forall j \text{ and } i < j \qquad (12.27)$$

$$\lambda_{zij}^- \geq \rho_{zij}^- M \qquad\qquad \forall i, \forall j \text{ and } i < j \qquad (12.28)$$

$$\rho_{xij}^+ + \rho_{xij}^- + \rho_{yij}^+ + \rho_{yij}^- + \rho_{zij}^+ + \rho_{zij}^- = 1 \quad \forall i, \forall j \text{ and } i < j \qquad (12.29)$$

4.4. The objective function of the model reformulation

The *nonintersection* conditions in the reformulated model are then expressed by constraints (11.16)–(11.21) and (11.23)–(11.28).[3] A region Q_{ij}, delimited by the constraints (11.16)–(11.21), is associated to each set of (feasible) values for the variables $\lambda_{xij}^+, \lambda_{xij}^-, \lambda_{yij}^+, \lambda_{yij}^-, \lambda_{zij}^+, \lambda_{zij}^-$ (for and $i < j$). The following *objective function* is introduced:

$$\max \sum_{i<j} (\lambda_{xij}^+ + \lambda_{xij}^- + \lambda_{yij}^+ + \lambda_{yij}^- + \lambda_{zij}^+ + \lambda_{zij}^-) \qquad (11.30)$$

For any (integer) optimal solution, the set $Q = \cup\{Q_{ij}, \ldots\}$ is 'minimal' (or equivalently the sum of terms $M_{xij}^+, M_{xij}^-, M_{yij}^+, M_{yij}^-, M_{zij}^+, M_{zij}^-$, appearing in the original model, is minimized).

Then we do not look for the optimal solution but the search is stopped when a first integer solution has been found. The *three-dimensional packing problem* formulated here is in fact solved when conditions (11.1), (11.2)–(11.4), (11.5)–(11.7), (11.19)–(11.21) and (11.22) hold.

V. THE OPTIMIZATION ENVIRONMENT

The presence of extra conditions is quite common in practice. With the (modeling) approach adopted, requirements such as the assigned position or the predefined orientation for some particular item may be treated immediately by fixing variables; other, more complicated, extra conditions may be taken into account by introducing additional constraints (provided that they can be expressed in terms of MIP).

When tackling difficult instances, heuristics are often useful to reduce the computational time. The heuristic algorithms introduce additional conditions on sub-sets of parallelepipeds and fix variables or properly modify the target function. Moreover, the necessity to evaluate and compare different options (corresponding to the additional conditions introduced, or representing particular design and operational choices) arises frequently.

In this context, an advanced modeling capability, as well that of visualizing (even during the numerical process) the various (final or intermediate) solutions is important. A dedicated optimization environment has been set up. It is based on the utilization of an algebraic modeler (EasyModeler, 1994), an optimizer (OSL, 1992) and a graphical system (IBM CATIA).

A two-way interface between the optimizer and the graphical system has been implemented. It allows friendly definition of the model input and visualization of the output, as well as interaction with the numerical elaboration (at any step of the iterative process) by 'suggesting' partial solutions (for example, fixing variables).

VI. COMPUTATIONAL RESULTS

An accommodation study case is reported here, as an application example. It consists of the (orthogonal) allocation of 20 small items (parallelepipeds) of different dimensions into a drawer (parallelepiped).

Any orthogonal rotation is admitted for almost all the items involved, while the remaining have a predefined orientation and no further extra conditions are imposed. About 80 percent of the volume is occupied.

A heuristic approach may be adopted for the above study case. It is described here in a simplified version:

- 15 items have been selected and the corresponding MIP model solved (that is a first integer solution found)
- sub-sets of ρ variables of the first integer solution have been fixed at the values attained
- the instances involving all the 20 items have been generated without the constraints made redundant by the fixed variables
- A further run has been performed.

In this way, only (all or part of) the nonintersection conditions, involving the first 15 items, are settled: just their relative positions have been identified, while they still may be moved or rotated if the rotations were admissible.

The original instance of 20 items (with no fixed variable) would approximately contain: 1200 continuous variables, 1300 binary variables and 2600 constraints running on IBM RISC S/6000 model 560.

With the above procedure, solutions such as that reported in Figure 11.3 (where domain D has been omitted) may be found in less than 1 minute of CPU time. However, as the approach is based on fixing variables, it may arbitrarily introduce infeasibilities (and consequently no solution will be found). A more sophisticated (even if computationally heavier) heuristics could be adopted: no variable is fixed, but proper terms are introduced into the target function, so that the ρ variables are oriented towards the values previously attained.

Even if this approach is generally quite advantageous, computational experience demonstrates that the elaboration time is strictly dependent on the specific instance, so that it may range from some CPU seconds to some minutes. The above example is quite representative of usual accommodation scenarios (at container level), even if generally the occupied volume is lower, for accessibility and operability reasons. The number of containers and large items per rack normally does not exceed 15 and the total number of racks is 16.

On the basis of the computational experience the volume availability is the most critical aspect. Below a threshold value (25 in our experiments), the number of parallelepipeds does not affect the running time. The impact of domain typology (dimensions, rates between the sides of each item, rates between items and domain sides) depends on specific

Figure 11.3   The equivalent solutions of the sample problem

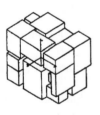
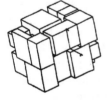

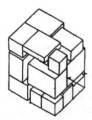

Table 11.2   MIP models dimensions

Items number	Variables number	0–1 variables number	Constraints number
15	~1450	~800	~1500
20	~2500	~1300	~3000
25	~3900	~2000	~4600

Table 11.3   Computational performances of problems with no extra constraints

Number of items	Volume filling rate	OSL default MIP CPU time in secs	Heuristic approach CPU time in secs
15	60	15	
15	65–75	60	
15	80–85	300	
20	60	30	
20	65–75	120	
20	80–85	900	60
25	60	60	
25	65–75	600	
25	80–85	1800	240

instance. Preassignment of items' position or orientation is easy to handle and often reduces the computational difficulties.

On the other hand the items' mass distribution heavily affects the static balancing. Tables 11.2 and 11.3 refer to models with no external conditions. The solution approach combining an OSL MIP preprocessor with default branch-and-bound parameters (IBM, 1992) appears the best strategy to tackle the problem family that we have tested.

VII. CONCLUSIONS

This chapter presents an advanced methodology to support cargo analytical integration in space engineering with reference, in particular, to the accommodation aspects. In order to reach satisfactory solutions the three-dimensional packing problems has been reformulated.

The optimization modeling approach allows us to introduce external conditions in several sample and real-world instances. A dedicated optimization environment seems appropriate to solve real problems.

Future activities could include the implementation of specific MIP strategies to speed up the numerical process, as well as the achievement of new heuristics. The extension of the whole methodology to different application areas, not only limited to the space environment, could be a further objective.

Notes

1. In 1988, Alenia Aerospazio was responsible for the design, development and integration of the ATV Cargo Carrier.
2. As one alternative, the logical condition (11.21) could be treated algorithmically, that is by introducing (non-standard) special ordered sets of variables (For the definition of the standard special ordered sets of variables *SOS1* and *SOS2* see, for instance, Williams, 1993.).
3. The following conditions are added with the scope of further reducing the computational difficulties (in fact, they make the LP-relaxed solutions 'closer' to the MIP ones):

$$\forall i < j \; \lambda_{xij}^+ + \lambda_{xij}^- \leq M, \lambda_{yij}^+ + \lambda_{yij}^- \leq M, \lambda_{zij}^+ + \lambda_{zij}^- \leq M.$$

References

Amadieu, P. (1997) 'The European Transfer Vehicle Mission and System Concept', 48th International Astronautical Federation Congress, Turin (October).

Boeing Defense & Space Group, Huntsville (Alabama) (1996) 'Cargo Planning, Analysis and Configuration System (CPACS) Software User Manual', S683-30014-1 (October).

Brinkley, R.H. *et al.* (1997) 'International Space Station: an Overview', 48th International Astronautical Federation Congress, Turin (October).

Daughtrey, R.S. *et al.* (1991) 'A Simulated Annealing Approach to 3-D Packing with Multiple Constraints', Boeing Huntsville AI Center, Huntsville (Alabama), Cosmic Program MFS-28700.

Dyckhoff, H., G. Scheithauer and J. Terno (1997) 'Cutting and Packing', in M. Dell'Amico *et al.* (eds), *Annotated Bibliographies in Combinatorial Optimization* (Chichester: John Wiley).

EasyModeler (1994) *IBM AIX EasyModeler, User Guide, Release 2.0*, SB13-5249, IBM Corporation, Rome.

Fasano, G. and R. Provera (1997) 'An Advanced Approach to Support the Cargo Analytical Integration of a Space Carrier', *Congresso Nazovale di Logistica*, Turin (October).

Nemhauser, G.L. and L.A. Wolsey (1988) *Integer and Combinatorial Optimization* (New York: John Wiley).

OSL (1992) *Optimization Subroutines Library, Release 2, Guide and Reference*, SC23-0519, IBM Corporation, Armonk.

Suhl, U.H. (1983) 'Solving Large Scale Mixed Integer Programs with Fixed Charge Variables', IBM Thomas J. Watson Research Center.

Williams, H.P. (1993) *Model Building in Mathematical Programming* (Chichester: John Wiley).

12 Horizontal Marketing: Optimized One-to-one Marketing[1]

Michael P. Haydock and Eric Bibelnieks

I. INTRODUCTION

As we approach the next century, direct marketers face a set of challenges that is far different than those the industry contended with during its explosive growth over the last twenty years. A quick synopsis of the current changing conditions brings this home. The percentage of households purchasing through the mail has been essentially flat since 1993. The consumer base is becoming increasingly diverse and individualistic. The average number of promotions received per household continues to climb. The cost of advertising (and in particular paper, postage, and ink) reached an all-time high in 1995 and continues to climb. Consumers increasingly demand services such as phone orders and expedited delivery as standard. New and often large traditional retail companies are adding direct marketing to their channel mix even as the industry consolidates. And consumers are presented with increasingly attractive alternatives to mail for the direct purchase of goods and services in their homes.

Direct marketers have responded to this tremendous amount of change in a variety of ways. Many direct marketers have invested in improving their targeted marketing capabilities, whether through the development of marketing databases, increasing the sophistication of their predictive models, or enhancing their current processes with leading-edge marketing tools, such as data mining. Often, these efforts have helped mitigate the impact of the challenges described above. That fact that the industry has not been able to improve its overall average response rate, however, indicates the improvements are at best keeping pace with the new challenges.

Rising challenges often present new opportunities to those who envision things from a different perspective. IBM believes that success in this new environment mandates new ideas, and Horizontal Marketing is one of our most exciting. This chapter explains why Horizontal

Marketing is a significant evolution from today's best practices. Among the things that set it apart are:

- making select decisions on an across-time (i.e. *horizontal*) basis that considers the range of upcoming promotions, not just the promotion at hand
- treating advertising as an investment instead of an expense
- borrowing from advancements in portfolio theory made by the securities industry
- adding linear and mixed-integer mathematics to the regression models used by industry leaders today and
- exploiting the dramatic decrease in cost and improvement in capability experienced by technology over the last several years.

The goal of this chapter is to provide an appreciation of the value Horizontal Marketing brings to optimizing customer relationships over time, a description of how a Horizontal Marketing application would be structured, and a brief description of the integer and linear programming models used to select promotions to mail to customers over a period of time vs. today's point-in-time efforts.

II. THE HORIZONTAL MARKETING APPROACH

The Horizontal Marketing approach was designed to create a 'discontinuity' relative to the traditional way that direct marketers have worked with predictive models. The central theme is that a one-to-one relationship can be developed over time between an individual customer and the firm. It seeks to develop the highest quality relationship with a customer while always keeping an eye out for the correct financial constraints of all customers as a group. The application looks at all feasible combinations of promotions and customers and selects the optimal relationship.

In our client consultations, we could see that most of the superb companies we work with are experiencing the baffling circular maze of spending more money on more media to reach customers less effectively. The regression and other models being used by our clients, although excellent efforts in their own right, continued to promote the same 'high RFM' (Recency, Frequency, and Monetary Value) customer set over and over again with very little variation. This saturates the best customers with promotions. Horizontal Marketing optimization models are designed to provide the best combination of relationship and affordability for each customer across time.

Horizontal Marketing scoring models in effect detect purchase 'patterns' the customer has exhibited in the past and simulates the customer reading each promotion in the mail plan, one promotion at a time. In essence, the customer goes on a shopping tour, one promotion at a time, until all promotions have been read and studied by our 'computational' customer. The models then judge what the reaction to each promotion may be and assign a 'fit' statistic that describes how well that particular promotion met that individual customers' needs with respect to the merchandise being offered, the season (or timing) being represented, and the type of promotion (or line of business) being presented.

This element of developing a relationship with the customer over time is the reason we named the technique 'Horizontal Marketing.' Also, consider a customer file as a huge spreadsheet with the rows representing customers and the columns representing promotions. Typically, direct marketers 'fill' this spreadsheet vertically by selecting who are the best customers for each promotion. Our technique allows us to fill the spreadsheet 'horizontally' by selecting what are the best promotions for each customer. By viewing each of the normal 'vertical' promotional events all at once and using mathematical optimization, we can determine the best overall treatment program (or relationship) for a complete planning period's worth of promotions. Furthermore, this technique allows to firm to shift its focus from optimizing individual events to optimizing the entire customer relationship.

These unique relationships are very valuable in that when we investigate individual customers we find that some relationships are worth more to the firm than others. We have described these relationships in a 'risk/return' paradigm where curves describe the relationship between the cost of a relationship treatment with the return on investment attained by that treatment. The challenge is to treat the relationship up to the point of diminishing returns (Figure 12.1). Past the point of diminishing returns, there is no financial gain attained under any level of relationship expenditure. When building the optimal diversified portfolio of customers, models use these curves and look for slope changes to make investment decisions.

2.1. The benefits of horizontal marketing

Horizontal Marketing offers a variety of benefits, both immediate and strategic. For many direct marketers, the most immediate benefit re-

Figure 12.1   A 'risk/return' curve

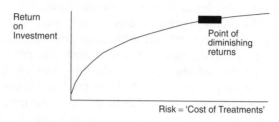

sults from a highly selective reduction in circulation. This can have a tremendous impact on business performance. Later we will present a the results of one mail test that yielded a 35% decrease in advertising expenditures coupled with a 66% increase in sales per advertising dollar.

More strategically, Horizontal Marketing better aligns the focus of your firm with your customers' needs. Today, direct marketers focus on maximizing the profitability of a specific promotion: 'we need 14% ROI and a minimum of $12 million in sales from the fall general merchandise catalog'. Horizontal Marketing shifts this focus to what can be done to better serve a homogeneous customer set across time: 'we need to provide more outerwear opportunities to customer set 23 via a specialized format'. This customer focus can also serve as a natural bridge between marketing and merchandising as they plan their promotion and offer strategies. For example, it provides a perspective on how to allocate advertising dollars across customer sets: 'we should remove $70000 in advertising from customer sets 23 and 42 and reallocate it to prospecting.'

2.2. Portfolio theory and direct marketing: investing advertising in your customers

The Horizontal Marketing approach can consistently out-perform leading targeted marketing methods based on single time-period discrete events. The primary theory that underlies the Horizontal Marketing concept is that customers of the firm can exist in portfolios. These portfolios can describe the optimal business relationship *across time* with that unique customer, and that these portfolios can be optimized to produce the maximum return on advertising, promotional or marketing investment, subject to a set of given corporate policy and investment constraints.

The theory's roots are based in the financial and investment industry and include the awareness that a diversified portfolio of assets (in this case, customers) can achieve maximum profits by quantifying the risk return relationships between the various asset types. Our methodology (efficiently) invests in each asset type up to the point where we obtain maximum return given a level of risk.

Horizontal Marketing defines return from an asset as the cash flow from a unique customer or group of customers with similar characteristics (i.e. a customer segment). Risk becomes the act of mailing a promotion to a customer given that the return on that particular mailing may be zero. Where most firms consider advertising dollars as an expense of doing business, the Horizontal Marketing approach treats advertising dollars as an investment. These advertising dollars can be invested across various groups of customers to maximize profitability over a defined time horizon, usually a fiscal year. These investments can also be constrained by any short-term corporate objectives that are required at the time of mailing a promotion (e.g. minimum acceptable program revenues or minimum circulation).

2.3. The confluence of Operations Research, database marketing, and information technology

Horizontal Marketing employs the goal-seeking mathematics typical of Operations Research problems. Using linear and mixed-integer programming techniques direct marketers can develop an economic model of their relationship to their customers. Included in this model are the nature of the assets/customer segments (such as purchase frequency, media affinity, seasonally, and incidence of returns) and the various aspects of risk (including promotional saturation, offer mix, and margins by media and offer type). Solving the economic model for the optimal promotional investment per customer is computationally demanding. For certain highly complex environments, the Horizontal Marketing approach could not be solved in a reasonable amount of time as recently as several years ago. Fortunately, the dramatic decreases in the cost of computing coupled with equally dramatic increases in computer performance now make Horizontal Marketing not only feasible but truly economically compelling.

III. THE HORIZONTAL MARKETING APPLICATION

In this section, we describe an example application, which was developed for a firm. This description of the application is very general while the models themselves are complex. In section IV, we will go into greater detail on the optimization models used to select the promotions a customer should be mailed. This section provides a basis and the background for the optimization models described in the next section. Please contact the authors if you desire a more detailed discussion.

The application is comprised of seven fundamental parts:

- database load from operational systems
- customer segmentation by investment profile
- budget allocation amongst customer segments
- promotion propensity scores and cannibalization matrix
- further customer segmentation by purchase patterns
- selection of optimal mail streams
- results uploaded to operational mail systems.

3.1. Database load from operational systems

The customer database that is used by the applications incorporates the customer characteristic data that are stored on each active customer in the firm's marketing database. This information primarily consists of behavioral data (purchase, returns, and potentially payment history), promotion history, and demographic data. It may also include data produced or derived from current scoring models.

The data is then typically transformed from the raw characteristic data that was stored in the 'data warehouse' by using a series of mathematical transformation applications created by the project scientists. These transformations are also stored in the warehouse or in a repository comprising relational tables. Data types could include customer characteristic data, time-dimensioned data, positional data, and external data.

3.2. Customer segmentation by investment profile

Strategic customer segmentation decomposes the customer base so that we have clusters of customers, which we call *asset classes*, that have identical or nearly investment profiles. Typically, this segmentation is performed with the goal of managing the resulting asset classes with investment strategies, which achieve shareholders' goals of revenue growth and maintaining/increasing profitability. For example, a firm may wish to invest money in an asset class which has a short history with the firm and low profitability, but the firm views as having a long-term potential as its members move to more profitable asset classes.

There are many methods to decompose customer data in such a way that they form these asset classes. Typically, to create asset classes we use classification tools which use patterns in the data, cluster analysis which clusters customers about 'centriod' customers, and/or business rules defined by the firm. We have found the most successful of these methods combine analytical processes with business rules provided by subject matter experts within the marketing department of the firm.

3.3. Budget allocation amongst customer segments

A cornerstone feature of the Horizontal Marketing approach is the ability to look at the decision process across many different dimensions of the firm. Those dimensions would include both customer group behaviors and multiple time periods. It seeks to answer the question: 'am I doing the optimal set of activities that maximizes my firm's profit while minimizing the cost of my firm's promotional budget?' This question has the dimension of time associated with it because the lifetime value of a customer is not determined in a single period but is similar to an annuity, which pays the firm in profits over many time periods.

When the marketing activities of the firm are envisioned as a long-term investment in a customer, the question becomes: 'in which customers should we invest, and how much?' Today, the mailing behavior at most firms reflect the effectiveness of the regression models that score a customer's propensity to buy. That score is developed with logic that tries to identify customers who bought from a previous promotion that was similar and tries to identify new customers who have characteristics similar to those who bought. Their scores in descending sequence rank the customers and a 'line' is drawn at the point where it is unprofitable to mail any deeper.

One problem with this approach is that recency (that is, the number of days since purchase) dominates the models so much that high-scoring customers end up receiving nearly every promotion. Some direct marketers are even getting requests from their good customer to promote them less often. While established direct marketers usually make mailing decisions that allow each promotion to meet its financial objectives, a different allocation of advertising resources would result if we shifted our focus from program profitability to customer profitability across time.

Investors use a technique called 'asset allocation' to determine the amount of dollars that should be spent in each investment family or asset class. The technique is a rigorous inspection of alternatives that produces a set of holdings over time that maximizes return at the minimum risk amount or minimizes risk subject to some minimum acceptable level of return. This concept is applied to the direct mailer firm's customer base in an effort to produce a 'portfolio' of optimal activities that maximizes return for the fewest advertising dollars while constraining the portfolio to meet certain corporate or mailing objectives. A variety of factors can enter into these models, ranging from the expected return on alternative uses of these funds through the minimum amount of advertising needed to sustain the desired customer relationship. These factors represent the 'marketing levers' the decision-maker can employ. In our applications, the 'risk/return' curves that describe diminishing returns are created for each asset class and the overall budget is spread across all asset classes and multiple planning periods using a linear program. The asset allocation application is considered a strategic model for multi-period scenarios. By contrast, the segmentation, scoring and picking models are tactical models.

Once the budget has been spread across all asset classes we are now ready to push the budget all the way down to the customer level. There are two steps before that process is performed. The promotion propensity scores and cannibalization matrix must be determined to provide the promotion to mail to customers given we know how much (i.e. the budget). This feature is described in the next sub-section.

3.4. Promotion propensity scores and cannibalization matrix

The promotion propensity models determine a 'fit' for a particular promotion for each individual customer. Typically, the direct marketing firm uses either multiple regression or logistic regression models to predict these 'fit' scores. We leverage these models and the firm's expertise to develop 'reward' scores, which predict the amount of gross profit the firm should expect from each customer when mailed a particular promotion. As an example, gross profit can be calculated as the gross sales minus returns, pre-ship cancels (both firm and customer originated), bad debt (if a credit granting institution), and cost of goods sold. The promotion propensity models are used for both profit-taking and prospecting activities, such as reactivation of existing customers and acquisition of new ones.

The central theme is to look over a planning period of promotions and make the right selections for that customer, while all customers in that asset class are competing for a scarce resource (budget dollars). Each customer and promotion combination is scored. The variable cost of the promotion being mailed can be deducted from this 'reward' score to provide a 'profitability' measure, which the optimization algorithm uses to select optimal mail streams. The budget allocation process's asset allocation algorithms set the overall budget. Strong scores and ability to deliver dollars on a specific promotion (mailing) are highly correlated. The strongest scores receive the most budget relative to others in their asset class.

The interaction of promotions is an extremely important component of Horizontal Marketing. 'Cannibalization' describes that portion of a promotion's sales that are consumed by a follow-on promotion mailed while the original promotion was still generating sales. And, conversely, a portion of the follow-on promotion's sales is consumed/cannibalized by the previous promotion. After close investigation we have seen that there are three forces that move this cannibalization effect: a merchandising component, a promotion type component, and a time component. The amount of cannibalization increases when similar promotions with similar merchandise are dropped fairly close together in time. The more time between the promotions, the less the cannibalization effect. Dissimilar promotions with unlike merchandise will have very small cannibalization effects. This holds true whether the promotions are mailed several weeks from each other or on the same mail date. We describe these interactions with a cannibalization matrix whose rows and columns represent promotions and each entry is interpreted as the row promotion's impact as a percentage drop in the column promotion's 'reward' score.

Cannibalization is similar to risk in that it can be either good or bad. Sometimes the combined effects of the promotions create a positive effect on revenues. Other combinations create a negative effect. It turns out that the key to understanding the effect on an individual customer is to evaluate the total promotional strategy. By evaluating the total strategy, an understanding of the incremental gains or losses due to multiple promotions can be developed. Horizontal Marketing uses this information to seek out the optimal strategy for each customer.

3.5. Further customer segmentation by purchase patterns

The sub-segmentation part of the application further decomposes the asset classes so that we have clusters of customers, called *micro classes*, within an asset class that have identical or nearly identical past purchase histories. This tool provides a superior targeting instrument that utilizes specific customer characteristics. It also allows the selection models to distinguish the amount of budget dollars distributed between different micro segments within an asset class.

Set theory dictates that any member of the set has identical characteristics or properties when compared to any other member of the set. Our segmentation techniques cut 'cells' extremely fine so that there are, for the most part, very few customers in each cell. One of the keys to the success of the Horizontal Marketing approach is the way we detect and handle very subtle differences in customer groups. We are able to use such fine-grained segmentation techniques by employing extremely fast computers that can manage hundreds of market segment cells vs. the tens of cells that most database or direct marketers are used to handling. Each cell within an asset class is truly handled in unique ways. This is best evidenced by looking at the *mailflags* (mail/no-mail indicators for a promotion) for customer groups and particularly the differences between cells that are near each other. Their mailflags are in a completely different pattern than their neighbors, while all members of a cell have nearly identical mailflags.

The mailflag file yields another interesting observation. When we backload promotions mailed under the predecessor promotional scoring strategy and attach them to the cells, there appears to be no apparent pattern within mailings to the same cell. This usually implies that the mailing behavior of the firm in the past is not consistent for customers who have nearly identical behaviors. This is an important observation because one of the benefits of the Horizontal Marketing approach is reducing the number of mailing combinations one could have by selecting only the mailing patterns that best fit the customers while delivering the maximum amount of profit. Segmenting our markets very fine is a key to this goal.

There are many clever methods to further decompose customer data in such a way that they form small granular micro classes within the strategic asset classes. Some tools available are optimization methods which draw cutting planes, neural networks which create cells, classification tools which use patterns in data to create cells, cluster analysis which clusters like customers into groups, and so on. All are good methods. We have found that most of these methods work well when you are not familiar with the customer base or the data attributes and want ways of segregating customers as a first cut at segmentation. Regression techniques to 'bin' data are very effective when the marketing departments have extensive knowledge about their data. IBM draws on several of these techniques to devise to optimum segmentation.

3.6. Selection of optimal mail streams

The promotion selection process is initiated when all the data on customers, promotions, budgets, cannibalization, and constraints has been read into the application. Potentially, this list could include promotion overrides, which typically represent up-front business rules for establishing certain mandatory mail/no-mail selection, and segment specific constraints, such as at least two promotions a month for the best customer asset class.

By considering all of a customer's reward scores and the cannibalization matrix, the application is able to select the future combination of mailings that best fits that customer's characteristics and budget constraints. The order of selection for these mailings is not dependent on what is available to mail on the next mail date (the current state of the industry) but rather what is available to mail over the next three weeks or six months or whatever time horizon the direct marketing firm chooses. This is really the essence of Horizontal Marketing. It sends the **right promotion at the right time**, minimizing the opportunity to send promotions of poor fit.

Depending on the length of the time horizon and the number of promotions per time period, the number of potential mailing combinations can be astronomically large. For example, consider 24 mailings over a four-month period to a customer list containing 15 million customers. This amounts to approximately 251 660 000 million customer/promotion mail stream combinations. By utilizing the recent advances of technology and mathematics, the selection process is able to determine what is the optimal mailing combination for each customer. In general, the application first finds a 'starter' set of candidate mailing combinations for each segment of customers. The starter sets are small sub-sets of the total number of potential mailing combinations. Each candidate mailing combination incorporates the reward scores, cannibalization effects, and budget constraints to determine the optimal combination. The second step of the application applies mathematical optimization techniques across all segments to determine the optimal mailing combination for each customer using the set of candidate mailing combinations associated with the customer's segment. We simultaneously consider the asset class budgets, the overall mailing depths, and any other user constraint that would be applied at the promotion or customer list level. Finally, the application checks to see if the solution is optimal with respect to the total number of potential mailing combinations. If it is not, additional mailing combinations are added to the candidate sets of each segment and the second step of the algorithm is repeated until the solution is optimal with respect to all potential mailing combinations.

After all customers, in all segments, in all asset classes have been assigned the optimal promotions and the entire mailflag file written, circulation counts are computed for each of the promotions as a result of all the mailing activity. In this way the user can quickly see how the application has recommended specific mailings for all promotions. This mailflag file contains a row for each customer in the database with all the current and future mailings recommended by the application for the planning period. The application is sensitive to slight changes in the database and will pick up purchases and non-responses by customers as they occur. The application should therefore be run on a regular basis.

The promotion propensity scoring and selection models are designed to accommodate a changing world. Promotions can be added or deleted easily. The application can also readily respond to changes in offer mix. Further, marketing decision-makers can use the models to run 'what-if' scenarios to simulate the impact on the mail plan or offer mix changes.

3.7. Results uploaded to operational mail systems

The final process is to pass the mailflag file with the customer number and mail/no-mail information for the mailing event being evaluated to the mailing systems. This is done with a query to the database to select the information on each customer and pass this information over the network to the mailing systems.

IV. AN EXAMPLE OF MAIL STREAM OPTIMIZATION WITH HORIZONTAL MARKETING

In this section, we describe an optimization model representative of an actual model developed for a large ($2 billion+ in annual revenues) direct-mail firm in the USA. We will introduce some terminology and notation to clarify the approach, and provide descriptions of the models.

4.1. Mail stream optimization (MSO) solution approach

A *mail plan* is a specific set of promotions that are to be mailed at specific times within a given time horizon. Let set $P = \{p,q\}$ contain the indices for the promotions being considered for selection. A *mail stream* is a specific sub-set of a mail plan and is generally associated with a customer or customer micro segment. A mail stream can be thought of as a vector of 0–1 mailflags sequenced by the promotion mail dates. A *mail stream expense* is the total advertising cost associated with all promotions in the stream. The expense is used in MSO models to enforce promotion size and segment budget constraints. A *mail stream reward* is used in MSO models to measure financial benefit of a mail stream and is derived from the expected reward that results from sending the mail stream to the customer or customer segment. The cannibalization matrix is denoted $S = \{s_{p,q}\}$ whose entry, $s_{p,q}$, represents the fraction by which the expected reward of promotion q is reduced by mailing promotion p.

The objective of the MSO models is, for a given set of customer segments (asset class and micro class) and a mail plan, to assign mail streams to customers that maximizes the total financial reward minus mail stream expenses and minus cannibalization effects, subject to promotion quantity and customer segment budgets.

In order to avoid considering an astronomical amount of mail streams, our solution approach reduces the number of customers who need mail streams assigned (through customer segmentation) and the number of mail streams to consider (through delayed Column Generation). Our delayed column approach is very similar to the approach used to solve the classic cutting stock problem. It is outlined in Figure 12.2.

4.2. Sample mail stream optimization formulations

In the representative models described below, we focused on getting immediate results (via a live mail test) by first implementing only MSG and MSS. Our ultimate goal is to use MSU to improve upon our results, once MSG and MSS are proven winners. Figure 12.3 illustrates our simplified approach. We use MSG to generate candidate mail streams for each micro class. Then MSS uses these candidate streams to maxi-

Figure 12.2   Delayed column approach

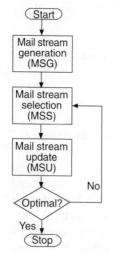

Mail stream generation (MSG) model
- constructs a set of candidate mail streams for each micro class

Mail stream selection (MSS) model
- selects the best mail streams in candidate sets for each micro class
- maximizes global objective
- enforces promotion quantity and budget constraints

Mail stream update (MSU) model
- updates MSS solutions with improved candidate mail streams

mize the total 'reward'/gross profit minus mail stream expense and cannibalization.

We will describe the models by the following characteristics:

- indices
- parameters
- variables
- constraints
- objective.

First consider the mail stream generation model. MSG's goal is to generate candidate mail streams (with the best possible net profit – i.e. gross profit minus expense) for each micro class j given the budget requirements for asset class k.

Indices:

$K = \{k\}$ the asset classes
$J^k = \{j\}$ the micro classes for asset class k
$P = \{p,q\}$ the promotions to be mailed within a specified time horizon

Figure 12.3 MSG and MSS

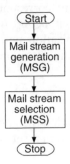

Parameters:

$R_p^{k,j}$ the expected 'reward'/gross profit from promotion p for customers in micro class j from asset class k

E_p the advertising expense for promotion p

$S_{p,q}$ the cannibalization of 'reward'/gross profit from promotion q by promotion p

$\underline{B}_k$ the lower bound on advertising expense to spend per customer from asset class k

$\overline{B}_k$ the upper bound on advertising expense to spend per customer from asset class k

Decision variable:

$y_p^{k,j}$ equals 1 when promotion p is mailed to customers in micro class j from asset class k, and equals 0 otherwise

Constraint:

Mail stream budget

$$\underline{B}_k \leq \sum_p E_p y_p^{k,j} \leq \overline{B}_k$$

Objective:

$$\max z = \sum_p (R_p^{k,j} - E_p) y_p^{k,j} - \sum_{p,q} R_p^{k,j} S_{p,q} y_p^{k,j} y_p^{k,j}$$

In order to generate n candidate mail streams for each micro class, the interval from $\underline{B}_k$ to $\overline{B}_k$ divided into n intervals and MSG was solved n times with the appropriate corresponding bounds. Furthermore, to convert the problem from being a 0–1 quadratic optimization problem to a 0–1 linear optimization model we linearized the quadratic term by using the following decision variable, constraints, and substitutions.

New decision variable:

$w_{p,q}^{k,j}$ equals 1 when promotion p and promotion q are mailed to customers in micro class j from asset class k, and equals 0 otherwise

Additional constraint:

Enforce linear variable to represent valid quadratic solutions

$$y_p^{k,j} + y_q^{k,j} - w_{p,q}^{k,j} \leq 1$$

New objective:

$$\max z = \sum_p (R_p^{k,j} - E_p) y_p^{k,j} - \sum_{p,q} R_p^{k,j} S_{p,q} w_{p,q}^{k,j}$$

Note that MSG is now linear. We solved MSG using the branch-and-bound algorithm provided by IBM's Optimization Subroutine Library. Now consider the mail stream selection model. MSS' goal is to select the best mail stream for each micro class j given the budget requirements for asset class k.

Indices:

$K = \{k\}$	the asset classes
$J^k = \{j\}$	the micro classes for asset class k
$P = \{p,q\}$	the promotions to be mailed within a specified time horizon
$M^{k,j} = \{m\}$	the candidate mail streams for asset class k, micro class j

Parameters:

$R_m^{k,j}$	the expected 'reward'/net profit from candidate mail stream m for customers in micro class j from asset class k
$F_m^{k,j}$	the advertising expense of sending candidate mail stream m for customers in micro class j from asset class k
$C^{k,j}$	the number of customers in micro class j from asset class k
$A_{p,m}^{k,j}$	equals 1 if promotion p is in candidate mail stream m for micro class j within asset class k

$\underline{Z}_p$ the lower bound on the quantity of promotions p to send
$\overline{Z}_p$ the upper bound on the quantity of promotions p to send
$\underline{B}_k$ the lower bound on advertising expense to spend per customer from asset class k
$\overline{B}_k$ the upper bound on advertising expense to spend per customer from asset class k

Decision variable:

$x_m^{k,j}$ the number of customers in micro class j from asset class k who receive candidate mail stream m

Constraint:

Promotion quantity

$$\underline{Z}_p \le \sum_{k,j,m \in M^{k,j}} A_{p,m}^{k,j} x_m^{k,j} \le \overline{Z}_p, \forall \, p \in P$$

Asset class budgets

$$C^{k,j} \underline{B}^k \le \sum_{k,j,m \in M^{k,j}} F_m^{k,j} x_m^{k,j} \le C^{k,j} \overline{B}^k, \forall \, k \in K$$

Micro class mailing requirement:

$$\sum_{m \in M^{k,j}} x_m^{k,j} = C^{k,j}, \forall \, k \in K, j \in J^k$$

Objective:

$$\sum_{k,j,m \in M^{k,j}} R_m^{k,j} x_m^{k,j}$$

MSS is a linear optimization problem that was solved with the simplex algorithm provided by IBM's Optimization Subroutine Library. We are optimizing the total 'reward', which can be interpreted as gross profit minus advertising expense and cannibalization, summed across all of the streams selected for mailing. We can control the quantity of promotions (this allows the direct marketing firm to meet postal requirements for quantity discounts), the advertising spent on each asset class, and ensure that every customer receives a mail stream. Note that the solution of MSS is continuous. In practice, however, we have noticed that the solution is almost always integer. When it is not we have developed a heuristic, which assigns the 'fractional' customer to the best mail stream of all the mail streams, selected by MSS for their micro class (within their asset class).

4.3. An analysis of a Horizontal Marketing mail test

Certain of our clients have conducted Horizontal Marketing validation tests. Control and experimental groups were selected randomly from the active candidates in participating firms' customer databases. We have run these tests with samples numbering well over 1 million customers. The objective of the validation testing was to compare promotional decisions made by the Horizontal Marketing model to the decisions made by the current models being used at the participating clients over a period of time.

The Horizontal Marketing application literally always looks backwards and forwards, evaluates the best set of options and activities of all possible combinations and executes a promotional (mailing) strategy, which presents itself in the form of a stream for each customer. Over the entire time period for which the mail test is run, the Horizontal Marketing application repeats this process every time the firm's current models run and only the promotions being selected by the firm's current models are selected from promotional stream recommended by the Horizontal Marketing application. This enables the Horizontal Marketing application to most closely mimic the firm's current operating environment and has access to the most recent information for making its decisions. This was how the Horizontal model was applied in the tests that were conducted. The Horizontal model was compared to the single event approach of the existing client methodologies, which look only at the current event to evaluate the best list. The current methodologies use the 'RFM' (Recency, Frequency, and Monetary Value) method of selection. The Horizontal Marketing approach also uses RFM components within its respective models also but introduces the dimension of time as well as other time-dependent variables (like cannibalization) in its decision-making process

Each model was given the actual advertising budget of the firm for that period. The Horizontal Marketing model uses a 'vent' in the budget allocation model that compares the financial return on mailing the 'next' promotion with the financial return on investing in a US government T-Bill. If the return on investing in the promotion is lower that the return on investing in the T-Bill, it chooses the T-Bill over the promotion. In that way, it always provides the user information on when the advertising budget has become more than the customer base can profitably absorb. This concept is illustrated in the first section of this work by the 'risk/return' curves. This is a direct result of saturation of the customer base by too many advertising dollars chasing too few customers. The Horizontal Marketing model will consequently not spend all of the advertising dollars given it.

The way the experiments were constructed allowed the appropriate samples to be taken for both the control and the experimental group. The control group was mailed using current selection methodology and the experimental group was tested using the Horizontal Marketing model.

The data for the control group was the actual mailing that was conducted with actual purchase results for the investment in advertising dollars. The members of the experimental group could be measured accurately when a promotion was sent and an actual purchase was made. The firm records for that event existed. What could not be measured accurately was when the Horizontal Marketing model mailed a customer that the firm did not initially mail. Since there was no resulting event of purchase or nonpurchase, a forecasting model was constructed that would measure the relationship between mailing and purchasing.

The forecasting model allows us to add actual purchases and forecasted purchases where the firms model did not choose to mail, but the Horizontal Marketing model did. Now we are prepared to compare the Horizontal Marketing model with the current methodologies. Table 12.1 reflects an example of a full test result.

The comparison of the control and experimental group details the promotions mailed in the experiment, the amount of advertising cost expended, the overall variable profit achieved by each model and the profit per advertising dollar produced by the models.

Table 12.1 represents the 'typical' results of the tests we have conducted. The key differences are in the advertising dollars spent between the two models and the profit obtained. The Horizontal Marketing approach is almost 65.52% more productive, in this example, comparing profit per advertising dollar. The Horizontal Marketing approach in effect spent almost 35.31% less in advertising and returned 7.07% more variable profit. Also, in every instance the Horizontal Marketing approach outperformed the firm's current model from a profit per advertising dollar perspective.

Table 12.1   Typical test results

Promotion	Firm's current approach			Horizontal marketing approach		
number	Advertising cost ($)	Variable profit ($)	Profit Per ad ($)	Advertising cost ($)	Variable profit ($)	Profit Per ad ($)
1	397 342	562 072	1.41	165 173	560 729	3.39
2	269 214	780 343	2.90	148 119	731 391	4.94
3	105 895	317 701	3.00	82 958	364 065	4.39
4	72 502	60 774	0.84	92 075	91 932	1.00
5	5984	5319	0.89	13 020	25 594	1.97
6	38 274	85 041	2.22	58 750	130 810	2.23
7	18 743	33 225	1.77	27 300	70 426	2.58
Total ($)	908 044	1 844 475	2.03	587 395	1 974 947	3.36

V. HORIZONTAL MARKETING BENEFITS

Horizontal Marketing is best suited to firms that have some of the following characteristics: a mail plan with over 10 programs per year and a variety of media or merchandise types; a mix of titles or businesses that mail into an overlapping customer set; a concern about the number of promotions being received by certain customer segments; a desire to better align the firm's strategic perspective with that of its customers.

One of Horizontal Marketing's primary advantages is the opportunity to achieve improved return from a firm's investment in advertising. These benefits will vary depending on the current intensity of your circulation strategy, the variety of media employed in your mail plan, the accuracy with which the horizontal strategy can be implemented given the current understanding of the customer base's preferences, the amount and type of data available, the variance in profit margins across media and offers, and others.

IBM can help companies develop a firm estimate of the advantages horizontal marketing might present both tactically and strategically. The formula below can be used as a rough estimate of the benefits available from one tactical area, changes in mailing behavior.

$$rB - \{[(1-a)rB] \times P \times m\} = \text{estimated pre-tax savings}$$

where:

a = accuracy, denoting the percentage of advertising dollars not spent given a reduction, r, in advertising investment that would have otherwise generated sales

B = the discretionary (or variable) advertising budget (advertising less such things as allocations for systems or salaries used to produce the advertisements, etc.)

m = the average profit margin on sales

P = the productivity of advertising dollars (average dollars of sales per dollar of advertising)

r = recommended or desired percentage reduction in the discretion ary advertising budget

This formula is only a rough estimate as other factors can contribute materially to the actual savings realized (e.g. the amount of advertising investment redirected into prospecting for new customers, etc.).

The process for migrating to Horizontal Marketing varies by firm. After gaining an understanding of your mail plan, offer set, available data, data structures, predictive modeling practices and business objectives, IBM will develop a customized plan tailored to your environment. In all cases, strong senior management involvement and a real enthusiasm for doing things in new ways are a must.

VI. CONCLUSION

Horizontal Marketing's across-time focus embodies the future of targeted customer relationship management. By combining many years of experience in database marketing and optimization mathematics, recent advancements in computing capabilities and economics, and exciting new ideas that treat advertising as an investment and customers as assets, Horizontal Marketing can substantially improve your business performance. Few opportunities hold as much potential, and IBM looks forward to working with leading direct marketers as we take targeted marketing into the twenty-first century.

Note

1. This document conveys concepts that are the intellectual property of IBM, the authors, and their associates.

Acknowledgments: The authors would like to acknowledge the following individuals from IBM who played important roles in developing the Horizontal Marketing concepts presented in this chapter Cindy Baune, Mark Bullock, Harlan Crowder, Wayne Kugel, Nancy Soderquist, and Steve Van Tassel.

13 The Early Classification Problem: Successive LP for Machine Learning

Tito A. Ciriani, Stefano Gliozzi and
Marina Russo

I. INTRODUCTION

Databased marketing, Horizontal Marketing and customer value management represent growing applications able to build one-to-one relationships with the customer.

Such marketing approaches address the final user instead of *product* or *brand* concepts; a consequence of this focus shift, is that the need to better understand the final users, and to group them in segments with the same behavior, to be addressed using the same marketing treatment, becomes more compelling.

We call segmentation the process of grouping elementary data (*instance of customers*) into *segments* having as far as possible a homogeneous behavior or relationship towards the company that owns the market. Segments of customer's data allow a company to implement consistent marketing strategies, to address specific customer needs, to anticipate possible attrition, to measure and value the market niches, and to identify new opportunities.

The practice of segmentation requires varied competencies mainly in statistics, data mining tools, and a deep knowledge of the explored market. In addition, the large amount of customers to be analyzed asks for a specific proficiency in handling Tera bytes' data. The author's experience suggests that to obtain a meaningful segmentation, the analyst should first derive from the raw data some efficient mathematical transformations, computing averages, slopes, deviations, and any kind of appropriate index; the segmentation process will then performed on these transformed data using one of the many available clustering/data mining algorithms.

Once the segments have been defined, their main characteristics interpreted, and eventually strategies implemented to deal the customer's segments, then the classification problem may arise.

With the classification problem we define the task of assigning an instance to a predefined segment. In general, owing to the nature of the clustering algorithms, the classification problems cannot be directly solved. The problem may be approached in two different ways. The first approach applies the logic of clastering algorithms. Most of the segmentation techniques implement a classification process that is usually efficient when the whole input data are available.

An alternate method is *trained* to solve the classification problem based only on the outcomes of the clustering. This second approach copes better with the lack of data in the original segmentation. We refer to this as to a *classification with incomplete data*.

The need for a classification with incomplete data may emerge directly from a business issue; for example, the segmentation could require, among other data, the *slope* or rate of customer expenses' change during the last two quarters. This data is by definition available only six months after the customer entered the database. But we may want to classify this customer instance – at least tentatively – well before six months from her entrance, to apply to her the best loyalty building strategy.

Similar situation happens when we want to know if a customer moves from a segment to another – maybe as a response to our marketing treatment. We may want to classify his instance again and to trace if he changed his segment in the most timely way. Then we need to use *incomplete* instances in order to test the effectiveness of a marketing strategy. Again, an educated estimation may be of great importance.

We follow an extension of machine learning method with linear discrimination (Mangasarian, Setiono and Wolberg, 1990). It allows classifying any data set for which a training set is available, regardless the segmentation-originating algorithm. In addition, our tests show that it supports slightly more reliable classifications than other algorithms based on neural network or decision trees. In particular, larger the missing data are the more efficient that method seems to be.

Section 2 briefly discusses the LP method for machine learning. Extension and EasyModeler (Ciriani and Gliozzi, 1994) implementation details follow. Then two real-life case results with comparison with other existing algorithms are presented. Some concluding remarks close the chapter.

II. MACHINE LEARNING WITH LINEAR DISCRIMINATION

Given any segmentation, we define as the *training process* the LP phase
that builds up the hyperplanes used to classify new objects. In addition,
let us define the variables we have available for the training process, as
the *vector of characteristics* of each observation. Let us also concentrate
on one training set of N observations. Among them we know that a sub-
set S has the characteristics that identify a specific segment. We need to
find two hyperplanes that separate the observations space in three parts
(if any) in such a way one can contain the segment's observations only,
another can not and a third one can contain both of them.

Figure 13.1 better explains this concept. Let us assume that the
observations have two characteristics.

The training set has eight observations and four of them (marked by
+) belong to the segment. The **a** line isolates one region of the plane
where the observations belong to the segment only. The **b** line identifies
another region of the plane with the observations that not belong to the
segment. These are the *classifying regions*. The observations, that can
belong or not to the segment, lie between the two lines.

The corresponding LP model is:

$$z = \min\,(b - a) \tag{13.1}$$

subject to:

$$\sum_i D_{ji} \cdot x_i \geq a \quad \forall j \in S \tag{13.2}$$

$$\sum_i D_{ji} \cdot x_i \leq b \quad \forall j \notin S \tag{13.3}$$

$$-1 \leq x_i \leq 1 \tag{13.4}$$

Figure 13.1   Separation of a segment

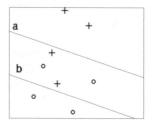

$$-\infty \le \mathbf{a}, \mathbf{b} \le \infty \qquad\qquad (13.5)$$

where:

> J is the training set (observations), indexed by j
> S is the sub-set of observations belonging to the segment
> I is the characteristics set, indexed by i
> D is the observation/characteristic matrix
> x_i are the unknown coefficients of the two separating hyperplanes
> a and b are the unknown right-hand-sides of the two separating hyperplanes.

Solving this model gives us a couple of parallel hyperplanes, with the following characteristics:

- when z is <0, the hyperplanes completely separate the element of S from the other observations
- when z is >0, the pair of hyperplanes defines three space regions
- when z is $=0$ all points lying on the hyperplanes are unpredictable. In addition this occurrence can address a pathological solution with all $x = 0$.

The $>a$ region where there may be only elements of S (if any), otherwise constraint (13.3) would be violated. The $<b$ region, where there can only be not-S elements (if any) otherwise constraint (13.2) would be violated. The observations belonging to such two *classifying regions* have been *classified*. The region $\le a$ and $\ge b$, that is in between, contains unclassified observations.

You should note also that the model could have a solution with a null x. To avoid the inconsistencies related to this situation we would like to be sure that the norm of x is strictly positive. If we assume that the norm of x as:

$$\|x\|_p = \left(\sum_i |x_i|^p\right)^{1/p}$$

we can define the infinite norm ($p = $ infinite) as the norm which approaches the maximum value of a x component. Thus we can ensure that we will not incur the pathological null x case just by fixing the x-norm to a predefined value, in our case 1. To achieve this goal we need to optimise $2 * I$ times the above model by fixing one at a time the x_i variables to -1 or 1 value.

Therefore, we need to solve the model in its pure form (without fixing any x_i) at the first iteration. If z is negative then the process stops since we already found the hyperplanes that discriminate the segments and we are ready to pass to another segment. Otherwise we will proceed through the $2 * I$ optimizations, and choose a pair of hyperplanes, according to some efficiency criteria.

A simple approach could be to select the hyperplane couple that leaves fewer observations in the *unclassified* space.

To build a complete *classifier* for the segment – i.e. a set of ordered hyperplane pairs able to discriminate the elements of the segment – the above procedure is iterated until the training set is empty, each time deleting all the classified observations. The set of ordered couple of hyperplanes identifies the *segment's classifier*, which can be used to classify an observation not belonging to the training set.

Each iteration checks if the observation belongs to one of the two classifying regions or remains unclassified. The unclassified observations will be considered for the next regions, and so forth. It may happen that the iterations terminate and the observation is still unclassified. In this case, we will be able to measure, at least, how *near* it was to being assigned to this segment. Mangasarian, Setiono and Wolberg (1990) provides the full theoretical details, including the *degeneracy*. Even if never occurred in their experiments, they define a *degenerate iteration* where no deletion of training set elements occurs. In our tests we have successfully removed degeneracy applying the solution proposed in the (1990) paper.

The algorithm described deals with a segment and its negation. We extended the use of the LP method to many segments by using the same model structure. Each model will have a different sub-set S, defining the different segments over the characteristics matrix D_{ji}.

Then the iterated model solution provides the best hyperplane couples that identify the segment's classifier. The set of all the *segment's classifiers* will then be our *multi-segment classifier*, *classifier* hereafter.

Overtraining, a well known phenomenon in machine learning, may occur when the classifier is extremely precise over the training set, at the expense of its *generalisation*, or the ability to capture the underlying characteristics of each segment. In order to reduce *overtraining*, we applied two more features: truncated classifier generation and multiple classifier generation with Taboo. Both techniques will be discussed in the next section.

III. IMPLEMENTATION

The dual formulation has been adopted in the implementation, to reduce the LP solution time

$$z = \max \left(\sum_i u_i - \sum_i v_i \right)$$

subject to:

$$\sum_{j \in S} D_{ji} \cdot A_j - \sum_{j \notin S} D_{ji} \cdot B_j + u_i + v_i = 0 \qquad \forall i$$

$$\sum_{j \in S} A_j = 1$$

$$\sum_{j \notin S} B_j = 1$$

$$-\infty \le u_i \le 0 \qquad\qquad\qquad \forall i$$

$$0 \le v_i \le \infty \qquad\qquad\qquad \forall i$$

where:

> J is the training set (observations), indexed by j
> S is the sub-set of observations belonging to the segment
> I is the characteristics set, indexed by i
> D is the observation/characteristic matrix
> u_i and v_i are the dual variables of coefficients of the two separating hyperplanes
> A and B are the dual variables of the right-hand sides of the two separating hyperplanes.

Since the number of characteristics (I) ranges from 10 to 100 and the number of observations of the training set may reach several millions, the immediate advantage comes from the reduced number of constraint to $I + 2$, while the number of variables increases to $J + 2 * I$. The opposite happens in the primal.

In fact, the dual model can allow an easy *Column Generation* approach (Chvàtal, 1983) when the training set expands to a very large number of observations. In addition, the optimization phase is faster for two main reasons: the model's matrix has few rows and each segment computes $2 * I$ optimizations each time starting from a feasible base with one coefficient of the objective function changed (to 1 or to -1).

The dual model has been implemented as a sub-routine of IBM EasyModeler Algebraic Language (EasyModeler, 1994). It also takes advantage of the exit routines to load the overall data only once.

Finally the data independence and the input matrix generation directly to the internal data structure of the IBM Optimization Subroutine Library (OSL, 1992), and the ability to keep the correspondence between the EasyModeler and the OSL structures through all the process, lead to an extremely compact and efficient model handling procedure.

The model generation sub-routine provided by EasyModeler solves the $1 + 2.I$ model instances by calling OSL at each model generation. The main loop drives the model generation and updates the training set at each hyperplane pair generation. The loop terminates when the remaining unclassified observations are lower than a selected value.

Such *truncated classifier generation* tends not to generate hyperplanes supported by too few observations (the *overtraining* problem). The classification accuracy may be affected, mainly for small-sized segments; in the test cases the threshold default value for unclassified observations is set to 10.

To improve classification accuracy, a mechanism similar to *Tabu search* is implemented to generate several classifiers (five in our test cases). During the search of each segment's classifier, the first hyperplanes pair is forced to be different from the one generated by previous classifiers. As a result the remaining training sets are (at least slightly) different and the overall classifier is therefore different.

This technique gives three distinct advantages:

1   it overcomes the disadvantage related to the *truncated classifier generation*, since the unclassified observations are unlikely to be the same for all generated classifiers
2   it increases the classifier's *confidence*, since it builds slightly different boundaries sets to characterize the same segment
3   at the classification time it computes the classifier's *confidence indicators*.

The main disadvantage of this technique is that it requires a training time as large as the number of classifiers generated.

The reported test cases required about five times the time to generate one classifier. Simple considerations can largely reduce the training time; the present implementation generates all five classifiers independently. Such a situation depends on the early training experiments that identify one classifier only. To improve the iterative process, after the first classifier has been computed, the initial model of the next classifier could get the advanced solution from the saved optimal solution of the previous one.

With several classifiers, the classification of new observation relies on a classifier's *voting* process. Each classifier, given the new observation, issues a *vote* for a segment. The *vote* is a function of the recognition and the search depth.

When the classifier assigns the new observation to a segment the vote value is 1, otherwise it is given 0.1. The assigned rate (vote) is scaled depending on what hyperplanes' couple the classification terminates (depth). The overall vote represents the sum of each classifier vote. The classification process assigns observations to the most voted segment.

Section IV presents the results of such approach tested on two real-world databases.

IV COMPUTATIONAL RESULTS

The MCG (multiple classifier generator) has been implemented from the technique described (Gliozzi, 1997) and it has been tested on real-world data using an IBM RS/6000 Model 560.

Table 13.1 Behavioral segmentation test of MCG: complete training set

Test 1 Complete training set	2% sample	5% sample	10% sample	20% sample
Training Set Accuracy	99.29%	99.37%	99.43%	99.40%
Newcomers' Accuracy	95.24%	96.37%	96.81%	97.22%
Overall Accuracy	95.32%	96.51%	97.08%	97.66%
Application Time	8'25"	9'24"	9'42"	12'38"
Training Time	49'53"	2h51'04"	7h36'04"	23h0'12"
Hyperplanes pairs number	433	851	1328	2454
Number of LP runs	13 207	26 353	46 822	107 182

The first data set (see Table 13.1) deals with a behavioral segmentation over a population of more than 280 000 customers. The segmentation, performed using the 'demographic clustering' algorithm of the Intelligent Miner IBM Product, leads to 34 segments, 15 of which accounted for more than 99 percent of the population. The segmentation refers to 25 numerical characteristics. The training set is formed by random samples of 2 percent, 5 percent, 10 percent, and 20 percent size extracted from such a data set. The training set contains elements from all the segments, but we computed the classifiers for only the 15 relevant segments. The entire population has been assigned to the segments using the samples' classifiers.

The classification test starts applying the classifiers to the training set (Training Set Accuracy). Subsequently we applied them to the overall population (Overall Accuracy). Finally we measured the assignment precision for the complete population of the sample, but excluding the training set observations (Newcomers' Accuracy).

Training Time seems to increase exponentially with sample's size, whilst the Application Time linearly depends on the classifier's hyperplanes. The Number of LP runs indicates how many times the LP model has been solved in order to find a satisfactory classifier, for all the data set's segments. The Hyperplanes pairs number refers to all the five multi-segment classifiers.

We then tested the ability of MCG to create reliable classifiers with highly incomplete information. To achieve this goal we considered the original segmentation from the Intelligent Miner Demographic Clustering method, which also gives as a result the ordering of the characteristics in terms of similarity within each segment. We chose to delete from the training set each segment's most homogeneous characteristic, thus we used in the training only 15 characteristics instead of the original 25 (Table 13.2).

Table 13.2 Behavioral segmentation test of MCG: incomplete training set

Test 1 Incomplete training set	2% sample	5% sample	10% sample	20% sample
Training Set Accuracy	99.17%	98.47%	98.16%	91.96%
Newcomers Accuracy	88.65%	90.03%	90.53%	85.19%
Overall Accuracy	88.86%	90.45%	91.30%	86.54%
Application Time	10'08"	15'25"	21'18"	37'48"
Training Time	1h15'36"	5h25'26"	16h52'05"	64h14m38"
Hyperplanes pairs number	1912	3296	5306	10089
Number of LP runs	47255	87154	145649	281366

The 86.54 percent value in the 20 percent sample seems to contradict the increasing accuracy related with the sample dimension. A reason can be found in the fact that within an incomplete test an enlarged sample generates many *identical* observations even if they belong to different segments. In such a case, a machine learning algorithm is clearly impaired, since it is not able to assign observations with identical characteristics to different segments.

The second data set (see Table 13.2) originated from a sample segmentation (less than 10 000 people) for credit scoring purposes. In this case, the population was to be assigned to 11 clusters. The original characteristics also contain a certain number of qualitative data, such as the Province code or the Industry Sector code. Each qualitative value transforms into a new binary variable and joins the training characteristics. After the merge their number increases to 34. The training set is a 10 percent, 20 percent, and 50 percent random sample, and the whole population (Table 13.3).

We also analyzed the classification's reliability for each single observation. For a given segment, we define the *reliability indicator* as the

Table 13.3   Credit-scoring segmentation test of MCG: complete training set

Test 2 Complete training set	10% sample	20% sample	50% sample	100% sample
Training Set Accuracy	100.00%	100.00%	100.00%	100.00%
Newcomers Accuracy	90.06%	91.14%	93.98%	–
Overall Accuracy	91.54%	92.91%	97.05%	100.00%
Application Time	21″	21″	23″	23″
Training Time	3′07″	11′12″	37′20″	2h0′43″
Hyperplanes pairs number	80	119	237	421
Number of LP runs	1598	2982	6721	11 430

Table 13.4   Reliability indicator of MCG over the behavioral segmentation test

	2%		5%		10%		20%	
Value	Obs	Correct	Obs	Correct	Obs	Correct	Obs	Correct
5	200 686	97%	196 497	98%	200 299	99%	207 721	99%
4	29 450	85%	24 998	86%	22 548	87%	19 467	87%
3	24 314	68%	19 925	69%	18 570	70%	16 004	72%
2	19 286	51%	16 195	51%	14 788	53%	15 585	47%
1	7761	41%	24 388	72%	7789	47%	4737	38%
0	1417	40%	741	28%	18 750	81%	19 230	82%

number of classifiers who positively voted for the observation belonging to the segment. Then five classifiers have been generated and used to classify the observations. Considering all segments, Table 13.4 shows how many observations have been successfully assigned to a classifier (Value).

We expect that the 5 value observations are almost always correctly classified, while 0 value observation is the result of a loose guess, since the observation lies in an 'unexplored' territory for all the classifiers. Because we know the segment to whom the test's observation (Obs) belong, we can reckon how many observations have been assigned to their own segment (Correct). The results relate to the classifiers generated from the first test of incomplete characteristics (Table 13.2).

Table 13.4 tells us that the observations with 0 indicator value (and in the 5 percent sample also with the 1 indicator value) seem to have larger correctness, when the sample size increases. This is due to the incomplete information of training set: as the sample increases, more observations are identical even if they belong to different segments. In the 2 percent sample, for instance, there are 12 identical observations, both belonging to segment **3** and **9**. The number of identical observations lying in different segments increases with the sample dimension , up to 604 in the 5 percent case, up to 2500 in the 20 percent case. Thus the training process is truncated when it is not able any longer to discriminate among segments. When the *unexplored* zone increases the classifier's accuracy decreases, but may occasionally dramatically increase whenever the *identical* observations are members of segments with very different cardinality. Whenever the cardinality of a segment is overwhelming (say, 70 percent), it may happen that the incomplete classification, by mere chance, attributes all the identical elements to that segment, achieving a great (and false) accuray (say, 70 percent).

V. LP VS. DECISION TREES AND NEURAL NETWORKS ALGORITHMS

For sake of completeness the LP approach has been compared with Decision Tree (DT) and Neural Network (NN) techniques. The benchmark refers to the same random samples and they are based on the state-of-the-art implementation of the algorithms, supported by IBM Intelligent Miner Product (Intelligent Miner, 1998). The final comments consider that the tests were run on a slower HW platform.

Table 13.5   LP vs. DT and NN: first test case with complete set

Test Case 1	Complete training set	LP	DT	NN
	Overall Population	95.32%	93.10%	94.38%
2% sample	Newcomers	95.24%	93.08%	94.36%
	Training Set	99.29%	94.32%	95.61%
	Overall Population	96.51%	94.45%	93.19%
5% sample	Newcomers	96.37%	94.40%	93.16%
	Training Set	99.37%	95.31%	93.78%
	Overall Population	97.08%	95.79%	93.22%
10% sample	Newcomers	96.81%	94.21%	93.19%
	Training Set	99.43%	97.60%	93.42%
	Overall Population	97.66%	96.61%	93.80%
20% sample	Newcomers	97.22%	96.45%	93.75%
	Training Set	99.40%	97.29%	94.02%

Table 13.6   LP vs. DT and NN: first test case with incomplete set

Test Case 1	Incomplete training set	LP	DT	NN
	Overall Population	88.86%	86.01%	80.38%
2% sample	Newcomers	88.65%	85.95%	80.34%
	Training Set	99.17%	89.14%	82.02%
	Overall Population	90.45%	88.42%	81.61%
5% sample	Newcomers	90.03%	88.32%	81.57%
	Training Set	98.47%	90.19%	82.39%
	Overall Population	91.30%	89.72%	81.76%
10% sample	Newcomers	90.53%	89.55%	81.71%
	Training Set	98.16%	91.18%	82.15%
	Overall Population	86.54%	90.65%	77.83%
20% sample	Newcomers	85.19%	90.37%	77.79%
	Training Set	91.96%	91.80%	78.04%

Table 13.5 reports the comparison in terms of accuracy of the three methods over the first test case. Table 13.6 shows the comparison results over the first test case with incomplete data and Table 13.7 refers to the second test case. In the first test set, with complete training, the LP-based algorithm seems to have a clear advantage over both the DT and the NN. In the second test set, the NN behaves slightly better in the first two samples, but is slightly better in the 50 percent, and fails in the overall population test.

In the training with incomplete data (which has more interesting applications) the LP outperforms the NN, and has a slight advantage in

Table 13.7   LP vs. DT and NN: second test case

Test Case 2	Complete training set	LP	DT	NN
	Overall Population	91.54%	90.34%	91.45%
10% sample	Newcomers	90.60%	90.24%	91.04%
	Training Set	100.00%	91.19%	95.16%
	Overall Population	92.91%	91.14%	92.16%
20% sample	Newcomers	91.14%	91.08%	91.56%
	Training Set	100.00%	91.41%	94.57%
	Overall Population	97.05%	92.91%	94.45%
50% sample	Newcomers	93.98%	92.47%	93.71%
	Training Set	100.00%	93.36%	95.28%
	Overall Population	100.00%	93.36%	88.01%
100% sample	Training Set	100.00%	93.36%	88.01%

the smaller samples over the DT. This advantage is lost when too many *identical observations for different segments* are in the sample.

On the other hand, even in the latter case, the advantage of having a high confidence indicator may be of practical use in several applications.

The performance evaluation is affected by different HW platforms used to run LP and DT and NN. Even with such a limitation, it is evident that the LP training phase is consistently slower than NN, and often the DT. In fact the DT and NN training terminates before the 2-hour run. On the other hand, during the classification phase, the LP performances are comparable to NN and considerably faster than DT.

VI. CONCLUSIONS AND FUTURE WORK

We have shown that the machine learning method with linear discrimination (Mangasarian, Setioni and Wolberg, 1990) can be easily extended to the multiple segment classification problem. The LP classifier generated by MCG can compete with DT and NN in terms of accuracy and ability to handle the amount of data. In addition the test cases show that that result holds also when the input data are qualitative, with no predefined ranking. The test cases show that LP is robust when the information is highly incomplete. The ability to produce a highly reliable classification trust indicator gives a distinct advantage in several practical cases.

The performances of the LP training are still unsatisfactory. The current MCG implementation seems not to be reliable when the training set becomes bigger than 50000 observations, and calls for more research in the area of a good criteria to choose among the $2 * I$ hyperplanes at each step.

The MCG approach is inherently parallel, since it requires the solution of several independent problems. In addition dynamic model update can be handled by EasyModeler sub-routines. On that base the future research will address three opportunities: a parallel computation; a Column Generation modeling for each hyperplanes pair, to allow MCG's training phase to deal with millions of observations; and different criteria to choose among hyperplanes at each step.

References

Chvàtal, V. (1983) *Linear Programming* (New York: W.H. Freeman).

Ciriani, T.A. and S. Gliozzi (1994) 'Algebraic Formulation of Mathematical Programming Models', in T.A. Ciriani and R.C. Leachman (eds), *Optimization in Industry, Volume* 2 (New York: Wiley).

EasyModeler (1994) *IBM AIX EasyModeler/6000, User Guide, Release 2.0*, SB13-5249, IBM Corporation, Rome.

Gliozzi, S. (1997) 'An LP Based Classification Algorithm,' *Technical Report*, IBM Consulting Group, Rome.

Intelligent Miner (1998) *Intelligent Miner for Data Version 2: Application Programming Interface and Utility Reference*, SH12-6326, IBM Corporation Armonk.

Mangasarian, O.L., R. Setiono and W.H. Wolberg (1990) 'Pattern Recognition via Linear Programming: Theory and Application to Medical Diagnosis', in T.F. Coleman and Y. Li (eds), *Large-Scale Numerical Optimization* (Philadelphia Society for Industrial and Applied Mathematics), 22–31.

OSL (1992) *Optimization Subroutine Library Release 2, Guide and Reference*, SC23-0519, IBM Corporation, Armonk.

14 Short-term Operation of an Electric-power System

Thomas Lekane and Jacques Gheury

I. INTRODUCTION

For many decades, electric power systems have been faced with one of the major problems of optimizing the short-term operation of their resources. This problem is known in the electricity supply industry as the unit commitment (UC) problem. It consists of determining both the hourly schedule and economic dispatch of each supply-side resource over a period ranging from 1 day to approximately 1 week. The schedule of a resource specifies its hourly status (on-line or off-line), when it starts up and shuts down while the dispatch defines the output level of each operating resource. The upcoming changes in the electricity industry towards the development of competitive electricity markets in many countries has increased the importance for companies to dispose of unit commitment software enabling them to systematically analyze and formulate optimal bidding strategies. The solution has to minimize the sum of all operating costs while satisfying the equipment and operating constraints of the resources. The study period extends from 1 to 7 days.

The UC problem may be formulated as a nonlinear mixed-integer program (MIP). The difficulty to find the optimal solution of large-scale real world problems comes from the following concurrent features:

- the nonconvexity of operating constraints for thermal units
- the time linking constraints of hydro resources
- the logical constraints.

Heuristic methods generally using unit priority lists were first proposed to solve the UC models (Happ *et al.*, 1971). Then, most of the software used different variants of dynamic programming to determine the operation of units (Pang *et al.*, 1981). From the 1980s, the most successful approaches have been based on the application of the Lagrangian-relaxation method (Xiaohong *et al.*, 1995; Batut and Sandrin 1990; Wang *et al.*, 1995). At present, genetic algorithms are being used to solve the UC problem (Orero and Irving, 1997).

The model considers a system including nuclear units, fossil-fired units and pumped storage plants. Operating considerations are taken into account – such as the operating reserve constraint, the minimum operating power and ramping rate constraints, the commitment constraints of the fossil-fired units, the operation of the nuclear plants at constant output levels, and the energy constraints associated with the pumped storage plants.

The chapter first presents a solution approach based on the Lagrangian-relaxation method and discusses the difficulties associated with finding out a primal feasible solution. Then, a new solution system which is being investigated in the frame of the ESPRIT project called MEMIPS is described. The approach is based on a Column Generation method which integrates dynamic programming to generate feasible power profiles for the resources and linear programming using EMOSL to find a set of profiles minimizing the total operating cost while respecting the system constraints. Heuristics are then applied to determine integer solutions.

II. FORMULATION OF THE MODEL

The objective of the UC problem is to minimize the sum of all operating costs of the thermal units over the study period while satisfying various equipment and operating constraints and taking advantage of possible power sales to neighboring systems. These costs comprise: the hourly proportional costs of fossil-fired and nuclear units, noload, fixed operating and maintenance costs of fossil-fired units, fixed and variable start-up costs of fossil-fired units and finally energy purchase costs. No cost is associated with the operation of pumped storage units. In the context of the Belgian system, two types of supply-side resources are considered in the model: thermal generation units including fossil-fired and nuclear units and pumped storage plants.

2.1. Notation

Sets are:

N Set of fossil-fired units
J Set of nuclear units
S Set of pumped storage plants

Ls Set of units of pumped storage plant s
T Set of hours
D Set of days in the study period
Td Set of hours of the day d

Indexes are:

i	Fossil-fired unit index	$i \in N$
j	Nuclear unit index	$j \in J$
s	Pumped storage plant index	$s \in S$
l	Pumped storage unit index	$l \in L_s$
t	Hour index	$t \in T, t \in T_d$
d	Day index	$d \in D$

Variables are:

pf_{it} Production of fossil-fired unit i at hour t; the sum of the production of all segments of unit at hour t

rf_{it} Operating reserve associated with fossil-fired unit i at hour t

uf_{it} Status of fossil-fired unit i status at hour t (1 = on-line, 0 = off-line)

xf_{it} Number of hours for which fossil-fired unit i has been on-line (if > 0) or off-line < 0) at hour t

pn_{jt} Production of nuclear unit j at hour t

rn_{jt} Operating reserve associated with nuclear unit j at hour t

zn_{jt} Number of hours for which nuclear unit j has been operating at the same output level at hour t

pp_{lt} Consumption of pumped storage unit l in the pumping mode at hour t

pt_{lt} Production of pumped storage unit l in the turbining mode at hour t

rh_{lt} Operating reserve associated with a hydro unit l at hour t

v_{st} Equivalent level of the upper reservoir of pumped storage plant s at the beginning of hour t

up_{lt} Pumping mode indicator of pumped storage unit l at hour t (0 = no, 1 = yes)

ut_{lt} Turbining mode indicator of pumped storage unit l at hour t (0 = no, 1 = yes)

mp_t System pumping mode indicator at hour t (0 = no, 1 = yes)

mt_t System turbining mode indicator at hour t (0 = no, 1 = yes)

Symbols with under-bar represent lower bounds while symbols with over-bar represent upper bounds of variables.

2.2. Resource modeling

Fossil-fired units

Cost characteristics The hourly cost characteristics of a fossil-fired unit consists of the sum of three terms, associated with the unit when it is on-line:

- a generation cost $C_i(pf_{it}) + CNL_i$, function of the generated power pf_{it} (see Figure 14.1)
- an hourly operating and maintenance cost $COMF_i$
- a start-up cost $CSU_i + CVSU_i(xf_{it})$ at each hour where the unit starts up (see Figure 14.2); an exponential function of the downtime duration.

Technical an operational characteristics A fossil-fired unit is subject to the following constraints:

- the maximum contribution of the unit to the operating reserve completion

$$rf_{it} = rf_i^o \qquad \text{if } uf_{it} = 0$$
$$0 \le rf_{it} \le \min(\overline{rf}_i, \overline{pf}_i - pf_{it}) \quad \text{if } uf_{it} = 1$$

Figure 14.1   Hourly generation cost curve

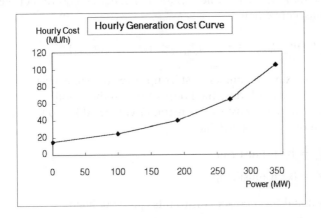

Figure 14.2   Start-up cost (MU)

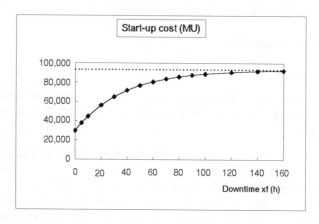

- the capacity limits of the unit

$$pf_{it} = 0 \qquad \text{if } uf_{it} = 0$$
$$\underline{pf}_i \leq pf_{it} \leq \overline{pf}_i \quad \text{if } uf_{it} = 1$$

- the minimum up- and down-time constraints force the value of uf_{it} if

$$0 < -xf_{i(t-1)} < MINDT_i$$
$$0 < xf_{i(t-1)} < MINUT_i$$

- the ramp rates limit the variation of generated power between two consecutive hours

$$\underline{GF}_i(uf_{it}, uf_{i(t-1)}) \leq pf_{it} - pf_{i(t-1)} \leq \overline{GF}_i(uf_{it}, uf_{i(t-1)})$$

- the maximum number of start-ups over the study period limits the variation of uf_{it} over the hours of the study period
- must run and must stop constraints can be added to force a unit to run or to be unavailable:

$$uf_{it} = 1 \quad \text{if must run}$$
$$uf_{it} = 0 \quad \text{if must stop}$$

- optional fixed power constraints can be defined

$$pf_{it} = cst, rf_{it} = 0, uf_{it} = 1$$

- initial conditions of the unit can be defined

$$xf_{i0}, pf_{i0}$$

Nuclear units

Cost characteristics The hourly cost characteristics of a nuclear unit consists of an hourly generation cost curve of one segment: CVN_j.

Technical and operational characteristics A nuclear unit is subject to the following constraints:

- the maximum contribution of the unit to the operating reserve completion

$$0 \leq rn_{jt} \leq \min(\overline{rn}_j, \overline{pn}_{jd} - pn_{jt}) \quad \text{if } pn_{jt} > 0$$
$$rn_{jt} = 0 \qquad\qquad\qquad \text{if } pn_{jt} = 0$$

- the capacity limits of the unit

$$\underline{pn}_{jd} \leq pn_{jt} \leq \overline{pn}_{jd} \quad t \in T_d$$

- the minimum output level duration and the maximum number of output level switching in a day force the value of zn_{jt} at some hours
- the ramp rates limit the variation of generated power between two consecutive hours

$$\underline{GN}_j \leq pn_{jt} - pn_{j(t-1)} \leq \overline{GN}_j \quad \text{if level switching allowed}$$
$$pn_{jt} = pn_{j(t-1)} \qquad\qquad \text{if no level switching allowed}$$

- optional fixed power constraints can be defined

$$pn_{jt} = cst, rn_{jt} = 0$$

- initial conditions of the unit must be defined:

$$zn_{j0}, pn_{j0}$$

Pumped storage hydro units

Plant characteristics   A pumped storage plant is characterized by the maximum capacity of its upper reservoir $VMAXS_s$.

Unit characteristics   In the pumping mode, power limitations are

$$pp_{lt} = 0 \quad \text{if } up_{lt} = 0$$
$$pp_{lt} = \overline{PP}_l \quad \text{if } up_{lt} = 1$$

In the turbining mode, power limitations are

$$pt_{lt} = 0 \quad \text{if } ut_{lt} = 0$$
$$\underline{PT}_l \le pt_{lt} \le \overline{PT}_l \quad \text{if } ut_{lt} = 1$$

Efficiencies of unit l in the pumping and turbining modes are given by η_l^p- and η_l^t respectively.

Operating mode constraints   All the pumped storage plants must operate in the same mode at each hour of the study period

$$mt_t + mp_t \le 1$$
$$ut_{lt} \le mt_t$$
$$up_{lt} \le mp_t$$

Plant operating constraints
- The level of the upper reservoir of plant s is constrained at hour t by

$$\underline{V}_{st} \le v_{st} \le \overline{V}_{st}$$

- The maximum overall power of plant s in pumping mode at hour t is constrained by

$$\sum_{l \in L_s} pp_{lt} \le \overline{PP}_{st} \text{ if } up_{lt} = 1$$

- The maximum overall power of plant s in turbining mode at hour t is constrained by

$$\sum_{l \in L_s} pt_{lt} \le \overline{PT}_{st} \text{ if } ut_{lt} = 1$$

Maximum operating reserve The maximum contribution of the plant to the operating reserve completion is defined as follows

$$rh_{lt} = \begin{cases} \overline{PT}_l - pt_{lt} & \text{if } ut_{lt} = 1 \\ \overline{PT}_l & \text{if } ut_{lt} = 0 \text{ and } up_{lt} = 0 \\ pp_{lt} & \text{if } up_{lt} = 1 \end{cases}$$

Energy conservation The evolution of the level of the upper reservoir of the plant is computed by the following formula

$$v_{s(t+1)} = v_{st} + \sum_{l \in L_s} \eta_l^p \cdot pp_{lt} - \sum_{l \in L_s} pt_{lt} / \eta_l^t$$

Initial and final conditions

$$v_{s1}, v_{s(T+1)}$$

2.3. System constraints

The system constraints include the demand and reserve requirement constraints at each hour of the study period:

- Demand constraints

$$\sum_{i \in N} pf_{it} + \sum_{j \in J} pn_{jt} + \sum_{s \in S} \left(\sum_{l \in L_s} (pt_{lt} - pp_{lt}) \right) = DEM_t$$

- Reserve constraints

$$\sum_{i \in N} rf_{it} + \sum_{j \in J} rn_{jt} + \sum_{s \in S} \left(\sum_{l \in L_s} rh_{lt} \right) \geq RRS_t$$

2.4. Objective function

The objective function to minimize represents the operating cost of the fossil-fired and nuclear units over the study period, as described in section 2.2.

$$\sum_{t \in T} \left(\sum_{i \in N} \{COMF_i \cdot uf_{it} + CNL_i \cdot uf_{it} + C_i(pf_{it}) + [CSU_i + CVSU_i(xf_{it})] \right.$$
$$\left. \cdot (1 - uf_{i(t-1)}) \cdot uf_{it} \} + \sum_{j \in J} CVN_j \cdot pn_{jt} \right)$$

III. LAGRANGIAN-RELAXATION APPROACH

In this approach, the demand and operating reserve constraints at each hour are relaxed. If λ and μ are the Lagrange multipliers associated to the demand and reserve constraints, respectively, the following dual problem can be written

$$
\begin{aligned}
L(\lambda,\mu) = \min_{\substack{uf,xf,pf,rf \\ pn,rn \\ pt,pp,rh}} \sum_{t=T} & \left\{ \sum_{i=N} (COMF_i \cdot uf_{it} + CNL_i \cdot uf_{it} + C_i(pf_{it})) \right. \\
& + [CSU_i + CVSU_i(xf_{it})] \cdot (1 - uf_{i(t-1)})) + \sum_{j \in J} CVN_j \cdot pn_{jt} \\
& + \lambda_t \cdot \left\{ +DEM_t - \sum_{i \in N} pf_{it} - \sum_{j \in J} pn_{jt} - \sum_{l \in L_s}(pt_{lt} - pp_{lt}) \right\} \\
& + u_t \cdot \left. \left\{ +RRS_t - \sum_{i \in N} rf_{it} - \sum_{j \in J} rn_{jt} - \sum_{s \in S}\sum_{l \in L_s} rh_{lt} \right\} \right\}
\end{aligned}
$$

subject to constraints defined here above for the fossil-fired units, the nuclear units and the pumped storage hydro units.

That can be written:

$$
\begin{aligned}
L(\lambda,\mu) = \min_{\substack{uf,xf,pf,rf \\ pn,rn \\ pt,pp,rh}} \sum_{i \in N}\sum_{t \in T} & \{ COMF_i \cdot uf_{it} + CNL_i \cdot uf_{it} + C_i(pf_{it}) \\
& + [CSU_i + CVSU_i(xf_{it})] \cdot (1 - uf_{i(t-1)}) \cdot uf_{it} - \lambda_t \cdot pf_{it} - \mu_t \cdot rf_{it} \} \\
& + \sum_{j \in J}\sum_{t \in T} \{ CVN_j \cdot pn_{jt} - \lambda_t \cdot pn_{jt} - \mu_t rn_{it} \} \\
& + \sum_{s \in S}\sum_{l \in L_s}\sum_{t \in T} \{ -\lambda_t(pt_{lt} - pp_{lt}) - \mu_t rh_{lt} \} + \sum_{t \in T} \{ \lambda_t \cdot DEM_t + \mu_t \cdot RRS_t \}
\end{aligned}
$$

As the Lagrangian function is separable in terms of resources, the UC problem can be decomposed into single sub-problems for each fossil-fired unit, nuclear unit, and pumped storage plant. For given values of λ and μ, each sub-problem can be independently solved in an efficient way.

3.1. Subproblem for each fossil-fired unit

The dual function is

$$l_i^{FF}(\lambda,\mu) = \min_{uf,xf,pf,rf} \sum_{t \in T} \{COMF_i \cdot uf_{it} + CNL_i \cdot uf_{it} + C_i(pf_{it})$$

$$+ [CSU_i + CVSU_i(xf_{it})] \cdot (1 - uf_{i(t-1)}) \cdot uf_{it} - \lambda_t \cdot pf_{it} - \mu_t \cdot rf_{it}\}$$

The state variables are:

- unit power
- number of consecutive hours with the same ON/OFF status
- number of start-ups since the beginning of the study period.

The transition status changes the unit power from the current hour to the next one. The operating power range is discretized, taking into account the heat rate curve segments and the ramping rates.

Some transitions are forbidden because of operating constraints (ramping rates, minimum up- and down-times, maximum number of start-ups over the study period, must run, must stops, fixed power).

3.2. Sub-problem for each nuclear unit

The dual function is

$$l_j^{NU}(\lambda,\mu) = \min_{pn,rn} \sum_{t \in T} \{CVN_j \cdot pn_{jt} - \lambda_t \cdot pn_{jt} - \mu_t rn_{jt}\}$$

The state variables are:

- unit power
- number of operating level switching since the beginning of the current day
- number of hours for which the unit has been operated at the current level.

The transition represents the change of unit level from the current hour to the next one. The operating power range is discretized, taking into account the ramping rates and the minimum and maximum daily operating powers. Some transitions are forbidden because of operating constraints (ramping rates, minimum duration of an output level, maximum number of output level switching in a day).

3.3. Sub-problem for each pumped storage plant

The dual function is

$$l_S^{PS}(\lambda,\mu) = \min_{pt,pp,rh} \sum_{l \in L_s} \sum_{t \in T} \{-\lambda_t(pt_{lt} - pp_{lt}) - \mu_t rh_{lt}\}$$

- The state variable is the upper reservoir energy level.
- The transition represents the change of upper reservoir energy level from the current hour to the next one, either by pumping (increasing the level) or turbining (decreasing the level).
- The upper reservoir capacity is discretized into a given number of energy intervals.
- When pumping, the constraint of integer number of pumps forbids transitions.
- When turbining, minimum and maximum operating power constraints of the turbines forbids transitions. Hourly constraints of minimum and maximum upper reservoir energy levels are taken into account.

3.4. Update of the Lagrangian multipliers

The λ and μ multipliers are updated using a heuristic based on a subgradient method. The advantage of this method is its ease of implementation and the rapid improvements of the solution in the first iterations. It reveals, however, convergence problems when hydro units such as pumped storage plants are considered; indeed, their hourly operation is very sensitive to small changes of the multipliers.

A major drawback of the Lagrangian-relaxation approach is that no primal feasible solution is obtained. So, the solution approach terminates with a heuristic tree search procedure to first determine a feasible unit commitment from the best solution to the dual problem, and-then to improve the solution.

This tree search method is based on the following concepts:

- the root node is the dual solution
- a child node is obtained by adding a unit up-time to its parent node
- three types of unit up-times are considered: base-loaded, cycling and peaking
- the candidate up-times are determined using dynamic programming

- at each node: evaluation of a lower bound of the solution cost
- a depth-first search is applied
- first objective: to find a feasible solution, and then to improve it
- each feasible solution is calculated using an economic dispatch model.

IV. SOLUTION APPROACH

Because of the difficulties already mentioned with the Lagrangian approach and the desire to extend the capabilities of the existing model by incorporating the representation of additional operating modes and system constraints, Tractebel is currently investigating a new solution method for the UC problem in the frame work of the MEMIPS project. This approach is based on a reformulation of the UC problem as a problem of determining the set of feasible thermal unit profiles which minimizes the total operation cost over the period studied while respecting the demand and reserve constraints as well as the pumped storage plant constraints. A feasible profile for a thermal unit is defined by both power and reserve profiles which specify its hourly generation and reserve contribution over time.

4.1. Principle

The principle of the proposed approach consists in generating a set of feasible thermal profiles and solving the MIP problem over the specified set of profiles. Since a huge number of thermal profiles exists, an approximate procedure is used to solve the MIP problem over a limited set of profiles. It comprises four main phases:

- **Phase 1**: to generate an initial set of feasible thermal profiles by using the solutions to the sub-problems found when solving the dual problem in the Lagrangian-relaxation approach.
- **Phase 2**: to solve the LP-relaxation of the MIP problem using a Column Generation technique.
- **Phase 3**: to solve the MIP problem using a relax and fix method if phase 2 does not lead to integer values for the binary variables.
- **Phase 4**: to solve the MIP problem obtained by fixing the binary variables associated with the thermal units at the values found as a result of phase 3.

In the Column Generation method, decomposing it into independent sub-problems for each thermal unit efficiently solves the sub-problem. Each sub-problem of the column generation method can be efficiently solved by decomposing it into independent sub-problems of lower level for each fossil-fired and nuclear unit, and using the algorithms of sections 3.1 and 3.2 developed by Tractebel.

4.2. Implementation

The solution approach is implemented in a single program by combining:

* modules of the existing unit commitment model written in FOR-TRAN by TEE to generate the thermal schedules
* the EMOSL library of DASH Associates (1997a, 1997b, 1998) to model and solve the LP and MIP problems
* routines written in FORTRAN and C programming languages to interface the thermal unit schedule generator, the thermal unit schedules data structure manager, and the EMOSL modules.

This program runs in the MS-Windows 95 and NT environments on a PC.

V. COMPUTATIONAL RESULTS

Test results are presented for the optimization of the weekly unit commitment of a power system with the following characteristics: 11 nuclear units, 79 fossil-fired units, 2 pumped storage plants, a peak load of 13 410 MW. Table 14.1 gives the solution cost of the Lagrangian-relaxation method original unit commitment model, as well as the cost at the end of phases 2–4 for the new solution approach.

The problem of phase 3 includes 2455 rows and 2465 columns, of which 1101 are binary variables, while the problem of phase 4 includes 71 769 rows and 21 192 columns, of which 672 are binary variables. Figure 14.3 shows the evolution of the solution cost during the solution process for the new approach; the five leftmost points are related to the final solution cost of each iteration of phase 2; the four next points represent the cost of the intermediate and final solutions of phase 3 while the rightmost point shows the final cost of the problem (phase 4).

The integrality gap is equal to 0.37%; it is defined by the following formula:

Table 14.1   Solution costs

Case	Cost	Elapsed time
Original Lagrangian-relaxation method (20 iterations)	1000.0	170″
Phase 1 (20 Lagrangian iterations)	–	85″
Phase 2 (4 iterations)	996.6	420″
Phase 3 (2 integer solutions)	1000.3	780″
Phase 4 (2 integer solutions)	997.6	220″

Figure 14.3   Evolution of solution cost

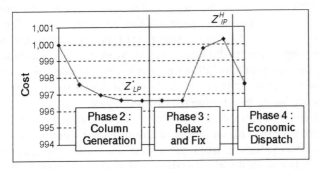

$$\text{Integrality gap} = \frac{Z_{iP}^H - Z_{LP}^*}{Z_{LP}^*}$$

where

Z_{iP}^H is the cost of the final integer solution of phase 3
Z_{LP}^* is the cost of the last iteration of the master problem of phase 2.

These values are identified in Figure 14.3.

The two programs were run on a PC Pentium 133 MHz. The elapsed times for the Lagrangian-relaxation method and the new approach are 170″ and 1505″, respectively, 95% of the CPU time of phase 2 is spent to solve the LP master problems.

VI. CONCLUSION

A new method is proposed to solve large-scale unit commitment (UC) problems. It is based on combining a Column Generation method, a relax and fix procedure, and a branch-and-bound method.

The new Lagrangian-MIP method aims to determine the values of the Lagrange multipliers by considering simultaneously the power profiles generated during all the iterations of the standard Lagrangian relaxation approach rather than the only ones generated at the previous iteration. A huge set of profiles should be taken into account to determine optimal values of the multipliers. In order to limit the set of profiles, a column generation method is used to iteratively build additional profiles improving the convergence to optimal multipliers

With respect to the direct recourse to a general branch-and-bound algorithm, this new approach, based on the application of Column Generation, takes advantage of the structure of the problem so that an efficient decomposition method can be used. Moreover, the technique for solving the sub-problems allows incorporating complex nonlinear modeling of power systems components which would be difficult to represent using linear programming (LP) formalism.

For small changes in the data, the possibility of freezing the values of some integer variables of the initial solution leads to finding a new solution not too different from the initial one. This capability of the approach is essential for industrial problems where implemented solutions cannot be completely reviewed in the short-term decision process.

The proposed approach is well suited to extending the current model formulation to deal with fuel switching, transmission constraints between areas, electricity sale and purchase contracts with other utilities, limitations on fuel supply and, environmental constraints. In order to accelerate the solution procedure, future work in the MEMIPS project will consider the improvement of the implementation of the Column Generation, the relax and fix heuristic, and the branching strategy in the branch-and-bound method.

References

Batut, J. and R.P. Sandrin (1990) 'New Software for the Generation Rescheduling in the Future EDF National Control Center', *Proceedings of the 10th PSCC*, Graz.

Dash Associates (1997a) *XPRESS-MP Optimization Subroutine Library XOSL, Reference Manual.*

Dash Associates (1997b) *XPRESS-MP User Guide and Reference Manual.*

Dash Associates (1998) *XPRESS-MP Entity Modeling and Optimization Library EMOSL, Reference Manual.*

Happ, H.H., R.C. Johnson and W.J. Wright (1971) 'Large Scale Unit Commitment Method and Results', *IEEE Transactions on PAS*, PAS-90, No. 3, 1373–83.

Lekane T (1997) *'Memips Esprit Project 20118* – Workpackage 2: Tractebel Model Analysis Report and Solution Approaches #2', DR 2.2.2, November.

Orero, S.O. and M.R. Irving (1997) Large Scale Unit Commitment using a Hybrid Genetic Algorithm', *Electrical Power and Energy Systems*, 19(1), 45–55.

Pang, C.K., G.B. Sheble and F. Albuyeh (1981) 'Evaluation of Dynamic Programming Based Methods and Multiple Area Representation for Thermal Unit Commitments', *IEEE Transactions on PAS*, PAS-100, No. 3, 1212–18.

Wang, S.J., S.M. Shahidehpour, D.S. Kirschen, S. Mokhtari and G.D. Irisarri (1995) 'Short-Term Generation Scheduling with Transmission and Environmental Constraints Using an Augmented Lagrangian-Relaxation', *IEEE Transactions on Power Systems*, 10(3), 1294–1301.

Xiaohong, G., P.B. Luh and L. Zhang (1995) 'Nonlinear Approximation Method in Lagrangian-Relaxation-Based Algorithms for Hydrothermal Scheduling', *IEEE Transactions on Power Systems*, 10(2), 772–8.

15 The Unit Commitment Problem

Samer Takriti

I. INTRODUCTION

Practically, all aspects of our modern lives involve electric power, from light bulbs and television sets, to hospitals and manufacturing. Although we are used to having power whenever we need it, the processes and systems involved in delivering electricity require careful planning and sophisticated mathematical and forecasting models. Therefore, it should come as no surprise that the electric power industry produces 12.7 trillion kilowatt-hours every year, consumes more than 2 billion tons of coal world-wide, and serves billions of people around the globe.

What distinguishes electric power from other sources of energy is that it is perishable and cannot be stored for future use. From an economic point of view, electric demand must be met the moment it arises. Similarly, on the production side, electric power must be consumed as soon as it is produced, otherwise, its production is pointless. As is the case with other commodities, electric power must be produced and transported to end users. To produce or generate electricity, utilities rely on generators which are fed with a particular form of energy – fuel, mechanical, or solar – that is transformed into electric power. To meet fluctuating electric demand, a utility uses a wide range of generating units. Some generators run at all times, while others are switched on and off as necessary. Generating units can vary in terms of fuel used, lead time, and generating capacity. These differences make the task of scheduling generating units, while meeting electric load, a challenging problem.

Once generated, electric power is carried across transmission lines to different distribution centers, which in turn distribute the power to consumers. Each time electric power is transmitted, some of it is lost owing to the resistance of the transmitting cables. To maintain a reliable service, a utility must be able to continuously generate enough electricity to meet the demand of its customers, plus the losses incurred in power transmission. Note that exceeding customers' needs increase the cost of operating the system and may cause power outages due to overloading some power lines. On the other hand, a generation shortage is definitely not tolerable. Social and economic losses that are incurred when electric demand is not met are not acceptable by consumers or government regulations. Given that power usage fluctuates continually, sophisticated tools are required to address the problem of controlling the load on each generator while meeting the demand.

To increase the reliability of an electric power system, transmission lines belonging to different utilities are connected to each other. These connections allow utilities to purchase power from each other when needed. For example, the impact of a supply shortage which occurs due to a generator failure can be lessened by distributing the burden of meeting demand with others. Of course, the utility that suffered a generator loss must start another unit as quickly as possible or take action to compensate for this loss.

To meet customers' demand, a planner predicts the future electric load, which is usually measured in megawatt-hours or MWHs. The forecast is given in the form of total load per hour for each period of the planning horizon – usually, 168 hours. For this reason, electric utilities employ forecasting experts and use advanced statistical techniques to predict the electric load on their systems. The main factor in determining future load is the weather – maximum and minimum temperature over each day, humidity, and cloud cover – at the different locations within a utility's service territory. In addition, a planner needs to take into account the effects of hour of the day, day of the week, holidays, and other events that might be relevant. Planners use load data from the previous week and from previous years to develop the load forecast.

Because of uncertainty in the weather forecast and in the electric load, utilities are required to have additional or excess generating capacity on-line. This is known as the spinning-reserve requirement, and states that the total generation capacity must exceed the predicted load by a certain factor at all times. The spinning-reserve requirement acts as a buffer against unusual events, such as a generating unit tripping down. The amount of excess capacity required varies from one utility to another. It is a function of the load profile and generation mix. For example, if a utility generates most of its power using hydro-electric units, one would expect the spinning reserves to be smaller than those of a utility which relies solely on nuclear units.

Using the load forecast, a utility schedules its generating units for the next 168 hours – that is, deciding when each generator should be on or off during the week. The schedule also indicates the load on each generator at each hour. A utility attempts to utilize a schedule that meets the load at a minimal cost. It is important to note that the resulting schedule must take into consideration all the constraints that are imposed by the characteristics of the generation system. The problem of finding a minimal-cost schedule for the generating units is known as the *unit commitment (UC) problem*. As the name indicates, it is the problem of deciding which generating units to commit at each time period of the planning horizon. To maintain high reliability in the electric system, a utility refines its load forecast frequently – every 8 hours – and solves a new unit commitment using the updated forecast.

Constructing a mathematical model for the unit commitment problem does not require full comprehension of the internal structure of a generator. However, it is helpful to understand how generators transform energy into electricity and the various types of generators that are used. We begin by describing the widely used steam-turbine generating unit. In general, a steam-turbine unit consists of a furnace, a water tank, and a turbine. Fuel burned in the furnace heats the water in the tank. As the water begins to boil, pressure from the generated steam drives the turbine-generator shaft, thus producing electricity. After cooling down, condensed steam (water) is pumped back into the tank so that it can be reheated. Part of the power generated is used to drive a cooling fan and to run the system's pump. In a fossil-fueled steam-generating unit, coal, gas, or oil can be used to fuel the furnace. Some units run on a single fuel while others burn more than one type of fuel, the latter being called a dual-fired unit. Often, natural gas is used first to start the system quickly, then the system switches to its primary fuel.

Understanding the previous description is sufficient to create a mathematical model. The output of a generator is a function of the heat generated in the boiler. The amount of energy fed into a generator, as a result of burning fuel, is usually measured in British thermal units or BTUs. The reader may think of a BTU as being the energy released by burning a match tip and a million BTUs as being the energy contained in 40 Kilograms of coal.

Figure 15.1 presents the relationship between the heat input to a generator and its electric-power output. The relationship is usually a second-degree polynomial. As expected, it is an increasing function – i.e. as we burn more fuel we expect to generate more electricity. One can use the curve of Figure 15.1 to compute the average heat input per MWH or the unit heat-rate. The unit heat-rate starts at a maximum rate (BTU/MWH) and declines as the load increases. This reflects the

Figure 15.1   Relationship between heat input and electric-power output

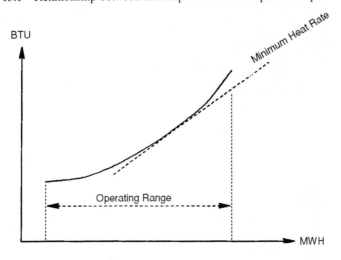

fact that the unit's efficiency increases as we increase the load. However, the efficiency usually declines after a certain load.

As there is a limit to the amount of fuel that can be burned in one hour, a generator has a maximum generation capacity. However, maintaining sufficient pressure to drive the turbine-shaft enforces a minimum limit on the amount of fuel that can be fed into a generator. This results in a minimum power output that is strictly positive. Note that during the process of switching a generating unit on, the water in the tank is far from boiling point. As a result, it may take a few hours before the generator starts producing electricity. The amount of energy consumed prior to beginning electric production is called the start-up energy. There might be a shut-down cost associated with cleaning the furnace and maintaining the unit between different runs. Finally, when a unit is switched on, it must remain on for a certain period of time. Also when off, it must stay idle for a given period of time. These two requirements are called the minimum up-time and minimum down-time constraints. The duration of the up- and down-time periods depends on the physical characteristics of the generating unit.

Other classes of generators use similar concepts to generate electricity. In the case of a nuclear unit, enriched uranium serves as the main source of energy. Water is either pressurized or boiled depending on the type of unit. As is the case with the fossil-fueled unit, a turbine transforms the kinetic energy of steam to mechanical energy that rotates the shaft. Another class of generators is that of turbines driven by gas pressure. For obvious reasons, they are called gas-turbine generators. Gases are usually produced in a high-pressure combustion chamber as a result of burning natural gas. These units have a small generating capacity and low efficiency. However, they can be dispatched quickly as the specific heat of gas is much lower than that of water. Finally, there are hydro-electric generating units. In this case, energy which is stored in water – water falls, rivers, or reservoirs – is utilized to rotate the turbine shaft. There is no need for a water tank or furnace to produce energy. Furthermore, hydro-electric units do not have a cost associated with them. When scheduling an electric system that has hydro units, the goal becomes to use them effectively so that the cost of thermal generation is minimized.

In terms of operation, generators can be classified into three categories: must-run or base units, cyclers or intermediate units, and peakers. Base units are the most efficient. Nuclear and steam-turbine generators are examples of base units. They run at a specific generation level all the time. One reason for this is that the time and energy required to start the system are costly, which makes switching a plant on and off economically infeasible. The load on base units is set to a certain level and kept at that level for a number of weeks. Base units cover most of the electric load on the system throughout the day – that is, their generation does not react to fluctuations in demand. As demand patterns change due to seasonal transitions, some of these units are switched on/off as necessary. The picture for cycling units is different. As the name indicates, cyclers are switched on and off depending on need. Of course, switching these units on and off must take the minimum up- and down-time requirements into consideration. The time required to start a cycler varies within a wide range, from 2 to 24 hours. As a result, careful planning is required to make sure that cyclers come on-line when needed. In some unusual cases, the load forecast misses the actual demand by a wide margin and the capacity of scheduled cyclers, in addition to base units, is not enough to cover the electric load. In this case, peaking units are called on. Their quick-start capability makes them a reliable safety buffer that can be used while other cycling units are being committed. Although peakers are costly to operate, their expense is justified as they help to maintain a reliable system that can deliver power under the majority of circumstances.

The generated electric power is carried across transmission lines to different points of usage or delivery. Power transmission and distribution is not relevant to our subject of interest – unit commitment; the interested reader may refer to Wood and Wollenberg (1996) as a starting point.

The rest of this chapter is organized as follows. Section II presents a mathematical formulation for the unit commitment problem. In section III, we discuss the Lagrangian solution approach and its related issues. In section IV, we extend the formulation of Section II to the stochastic case. Section V describes an integer programming formulation for refining the solution obtained from solving the dual. In section V, we apply our approach to the generating system of an electric utility. Section VI gives some numerical results, and section VII concludes.

II. PROBLEM DESCRIPTION

The following describes the input parameters and decision variables of the unit commitment model. The reader should be able to link the mathematical model with the physical description of section I. To fix notations, we assume that there are I generating units and that the duration of the study horizon is T periods. The state of a generating unit, i, at a time period, t, is represented by the 0–1 variable, u_t^i. A unit is on at time t if $u_t^i = 1$, and off if $u_t^i = 0$. The power output level at which unit i operates during a period, t, is $x_t^i \geq 0$. The cost of generating x_t^i MWH of electricity on i is $g_i(x_t^i)$. The function g_i is computed by multiplying the curve of Figure 15.1 by the cost of used fuel. Although multiple fuels may be used on the same unit, the average cost of the fuel used is easy to estimate. The function g_i is quadratic and convex. To switch a unit on, a start-up cost is incurred. Similarly, there might be a shut-down cost that represents the different expenses incurred when a unit is switched off. We use the function $h_i(u_{t-1}^i, u_t^i)$ to represent these expenses where $h_i(0,0) = h_i(1,1) = 0$. As a result, the cost of operating i at time t is $g_i(x_t^i)u_t^i + h_i(u_{t-1}^i, u_t^i)$. Note that if unit i is off, then its operating cost is 0. The minimum and maximum operating levels for each unit, i, are q_i and Q_i, respectively. When a unit i is switched on, it has to remain on for a certain number of periods, L_i. Similarly, if i is switched off it has to remain in that state for l_i periods.

The previous information is needed to describe the generation – supply-side of the picture. From the demand view, we assume the knowledge of the total electric load, d_t, at each period, $t = 1, \ldots, T$, of the planning horizon. We also assume that the amount of spinning reserves required at any period, t, is known. We denote the excess capacity required by r_t. That is, at any time period t, the utility must have a total generation capacity of no less that $d_t + r_t$.

Then, the unit commitment problem is to minimize the total cost of operating the electric system

$$\sum_{i=1}^{I} \sum_{t=1}^{T} [g_i(x_t^i)u_t^i + h_i(u_{t-1}^i, u_t^i)]$$

where u_0^i represents the initial state of generator i. The minimization is subject to

1 Meeting the expected electric load at each time period $\sum_{i=1}^{I} x_t^i \geq d_t$, $t = 1, \ldots, T$.
2 Having enough generating capacity to meet the spinning-reserve requirement, $\sum_{i=1}^{I} Q_i u_t^i \geq d_t + r_t$, $t = 1, \ldots, T$.
3 Respecting the generating limits of every generator at each time period of the planning horizon – i.e. $q_i u_t^i \leq x_t^i \leq Q_i u_t^i$, $t = 1, \ldots, T$, $i = 1, \ldots, I$.
4 Fulfilling the minimum up- and down-time requirements. Using the binary variables u_t^i, the minimum up requirement can be written as

$$u_t^i - u_{t-1}^i \leq u_\tau^i, \tau = t + 1, \ldots, \min\{t + L_i - 1, T\}, t = 2, \ldots, T.$$

Similarly, the minimum down-time requirement is written as

$$u_{t-1}^i - u_t^i \leq 1 - u_\tau^i, \tau = t + 1, \ldots, \min\{t + l_i - 1, T\},$$
$$t = 2, \ldots, T.$$

To take the initial conditions into consideration, additional constraints may be enforced on u_t^i. For example, if $L_i = 3$ and a unit has been on for 2 periods at the beginning of the horizon, then we must add the constraint $u_1^i = 1$.

To simplify notations, we denote the vector $x_t^i, t = 1, \ldots, T$, by x^i and the vector $u_t^i, t = 1, \ldots, T$, by u^i. We write the constraints

$$q_i u_t^i \leq x_t^i \leq Q_i u_t^i, t = 1, \ldots, T, \text{ as } x^i \in X_i(u^i) \text{ or } x \in X(u).$$

Furthermore, we denote the minimum up- and down-time constraints and the associated initial conditions by $u^i \in U^i$ or $u \in U$. As a result, the mathematical formulation of the unit commitment problem is

$$\left\{ \min_{u,x} \sum_{i=1}^{I} \sum_{t=1}^{T} [g_i(x_t^i)u_t^i + h_i(u_{t-1}^i, u_t^i)] : \sum_{i=1}^{I} x_t^i \right.$$

$$\left. \geq d_t, \sum_{i=1}^{I} Q_i u_t^i \geq d_t + r_t, t = 1, \ldots, T, u \in U, x \in X(u) \right\} \qquad (15.1)$$

The mathematical program in (15.1) is a mixed-integer program (MIP). Its size is determined by I and T. In practice, the number of generating units, I, can range anywhere between 10 and 200. The study horizon is usually set to 168 hours.

Among the approaches proposed to solve (15.1) are dynamic programming (Hobbs *et al.*, 1987) and branch-and-bound (Cohen and Yoshimura, 1983). Both of these techniques fail to solve practical unit commitment problems due to the large size of the search space. Priority lists (Lee, 1988) is a heuristic approach that ranks the generating units and commits them in the order of their ranking. This approach is quick, but results in approximate solutions that may be far from optimal. Simulated annealing (Zhuang and Galiana, 1990) and genetic algorithms (Kazarlis *et al.*, 1996) have also been tried, but the results were not encouraging. Given that these heuristic methods relax the constraints and incorporate them into the objective function, the solution produced at termination is often infeasible. The reason for this is that minimum up- and down-time constraints are difficult to satisfy if relaxed.

Instead of solving the unit commitment problem directly, one can relax the demand and spinning-reserve constraints. The advantage of this decomposition is that the mathematical program of (15.1) is decoupled into smaller single-generator problems. Each generator has its own optimization problem which can be solved using dynamic programming. As a result, state-dependent constraints, such as minimum up- and minimum down-time, can be handled easily. Another advantage of this approach is that it replaces a difficult mixed-integer program with that of maximizing the concave Lagrangian. However, the Lagrangian in this case is non-smooth which causes problems in updating the multipliers. Furthermore, an optimal solution for the Lagrangian may not have a corresponding primal feasible solution. Both of these problems need to be addressed if the dual problem is to be solved.

Muckstadt and Koenig (1977) seem to have been the first to suggest decomposing (15.1) by relaxing constraints that couple the generating units. The optimization problem of each generating unit is modeled as a dynamic program with T stages and 2 states – on and off – at each stage. The dynamic program is solved recursively, and the state of each generator, u_t^i, is determined in addition to its generation, x_t^i, at each period. The dual multipliers are then updated using a sub-gradient approach and the Lagrangian is re-evaluated. The procedure is repeated until a set of optimal dual multipliers is found. The value of the Lagrangian provides a lower bound on the objective function of (15.1). The backbone of the approach suggested by the authors is branch-and-bound. Depending on the feasibility of the resulting solution with respect to the reserve constraints, a variable u_t^i is selected and two new nodes – problems – are created. The first corresponds to $u_t^i = 1$ and the second corresponds to $u_t^i = 0$. A new problem is then selected from the list of active nodes. The process is repeated until all active nodes are exhausted.

For a practical system, the number of nodes in the search tree is expected to be large, which prohibits the use of such an approach. Muckstadt and Koenig observe this difficulty even with their relatively-small test problems, $T = 12$ and $I = 15$, but note that branch-and-bound 'tended to uncover good solutions early in the search through the tree. The remainder of the search was used to verify the goodness of this solution'. It is this remark that encouraged most researchers to find an optimal set of multipliers and avoid performing branch-and-bound to refine the solution or show its optimality. As a matter of fact, most subsequent approaches for solving the dual of (15.1) differ only in the choice of a step size and the heuristic adopted for enforcing feasibility. Indeed, Bertsekas *et al.* (1983) show that under the assumption that $I > 2T$, the duality gap corresponding to the previous relaxation is bounded above by a constant that depends on T and the physical properties – g_i and h_i – of the generating units.

Due to the non-differentiability of the Lagrangian, gradient information is not available at all points. As a result, most solution methods for the dual problem use a sub-gradient direction for optimizing in the dual space. The resulting methods often suffer from slow convergence. Lately, and due to advances in computer hardware and software, the use of numerically intensive methods in the hope of achieving better solutions and smoother convergence has become possible. Möller and Römisch (1995) use the bundle trust-region method of Schramm and Zowe (1992). The method takes reasonable steps while guaranteeing that the dual function is strictly increasing. If the current sub-gradient direction does not provide sufficient increase in the objective function, the search procedure tries to build an approximation for the dual at the current point by collecting a bundle of sub-gradients. As the reader may have noticed, this approach requires a large number of evaluations for the dual function. The authors solve problems with $I = 100$ and $T = 168$ and report encouraging results.

In the following section, we describe the Lagrangian-relaxation approach that is widely used for solving the unit commitment. We suggest a new mechanism for updating the dual multipliers that works well with the generating systems that we experimented with. The goal of our discussion is to provide the reader with a better understanding of how we solve the dual. This discussion will be helpful for introducing the refinement approach of section V.

III. SOLVING THE DUAL PROBLEM

As mentioned earlier, the mathematical program of (15.1) is solved by relaxing the demand and spinning-reserve constraints. To do so, we associate the dual multipliers $\lambda_t \geq 0$ and $\mu_t \geq 0$ with each of the constraints $\Sigma_{i=1}^{I} x_t^i \geq d_t$ and $\Sigma_{i=1}^{I} Q_i u_t^i \geq d_t + r_t$, respectively. The objective function becomes

$$\sum_{i=1}^{I} \sum_{t=1}^{T} [g_i(x_t^i)u_t^i + h_i(u_{t-1}^i, u_t^i) - \lambda_t x_t^i - \mu_t Q_i u_t^i]$$
$$+ \sum_{t=1}^{T} [\lambda_t d_t + \mu_t(d_t + r_t)]$$

To find an optimal set of multipliers λ and μ, we maximize the Lagrangian

$$\mathcal{L}(\lambda, \mu) = \sum_{i=1}^{I} \mathcal{L}_i(\lambda, \mu) + \sum_{t=1}^{T} [\lambda_t d_t + \mu_t(d_t + r_t)]$$

over the feasible region $\lambda \geq 0$ and $\mu \geq 0$, where

$$\mathcal{L}_i(\lambda, \mu) = \left\{ \min_{u,x} \sum_{t=1}^{T} [g_i(x_t^i)u_t^i + h_i(u_{t-1}^i, u_t^i) - \lambda_t x_t^i - \mu_t Q_i u_t^i] : \right.$$
$$\left. u^i \in U_i, x^i \in X_i(u^i) \right\}.$$

One can think of the Lagrangian iterative process as an auction in which prices for power, λ_t, and for reserves, μ_t, are offered for each time period of the planning horizon. Each generator responds to these prices (independently of other generating entities) through the decisions u_t^i and x_t^i. The bidding process is repeated in at attempt to force market equilibrium

$$\lambda_t \left(d_t - \sum_{i=1}^I x_t^i \right) = 0 \text{ and } \mu_t \left(d_t + r_t - \sum_{i=1}^I Q_i u_t^i \right) = 0.$$

It is expected that λ_t and μ_t will be positive. Therefore, the dual iterations aim at satisfying the demand and spinning-reserve constraints as equalities. This may not be possible due to the existence of a duality gap.

In the following sub-sections, we describe how to evaluate $\mathcal{L}_i\,(\lambda, \mu)$ using dynamic programming, how to initialize the multipliers and update them, and how to recover a primal feasible solution.

3.1. Dynamic programming

To evaluate $\mathcal{L}_i(\lambda, \mu)$, we use dynamic programming. There are T stages which represent the planning horizon. Each stage has $L_i + l_i$ states. The first L_i states correspond to the generator being on while the states $L_i + 1$ to $L_i + l_i$ correspond to the generator being off. If a generator has been on for less than L_i periods at t, then the only feasible decision is to keep it on so that the minimum up-time constraint is not violated. If a generator has been on for L_i periods or more, then there are two possible decisions: keep it on or switch it off. Similarly, if a generator has been off for less than l_i periods – i.e. it is in any of the states $L_i + 1, \ldots, L_i + l_i - 1$, then it must remain off. Finally, if it is in state $L_i + l_i$, then a generator can be kept off or switched on. We denote the minimum cost-to-go from the beginning of period t until the end of the planning horizon by $\mathcal{F}_i$. In other words, $\mathcal{F}_i(t,j)$ represents the minimum cost-to-go from the beginning of t given that the generator is in state j. The cost-to-go at time $T + 1$ is set to 0. Starting from $T + 1$, we move recursively backward using the optimality principle until we reach $t = 1$. Here is a sketch of the algorithm:

- **Initialization**. Set $\mathcal{F}_i(T + 1, j) \leftarrow 0, j = 1, \ldots, L_i + l_i$. Set $t \leftarrow T + 1$. Define $\bar{x}_t^i$ to be an optimal solution to $\{\min g_i(x_t^i) - \lambda_t x_t^i : q_i \leq x_t^i \leq Q_i\}$, $t = 1, \ldots, T$.
- **General step**
  1 Use $\mathcal{F}_i(t,j), j = 1, \ldots, L_i + l_i$, to evaluate $\mathcal{F}_i(t - 1, j), j = 1, \ldots, L_i + l_i$, as follows
     - If $j < L_i$, set $\mathcal{F}_i(t - 1, j) = \mathcal{F}_i(t, j + 1) + g_i(\bar{x}_t^i) - \lambda_t \bar{x}_t^i - \mu_t Q_i$ and $\mathcal{P}_i(t - 1, j) = j + 1$.

- If $j = L_i$, then if $\mathcal{F}_i(t,j) + g_i(\bar{x}_t^i) - \lambda_t\bar{x}_t^i - \mu_t Q_i < \mathcal{F}_i(t,j+1)$, set $\mathcal{F}_i(t-1,j) = \mathcal{F}_i(t,j) + g_i(\bar{x}_t^i) - \lambda_t\bar{x}_t^i - \mu_t Q_i$ and $\mathcal{P}_i(t-1,j) = j$; otherwise, set $\mathcal{F}_i(t-1,j) = \mathcal{F}_i(t,j+1)$ and $\mathcal{P}_i(t-1,j) = j + 1$.

- If $L_i < j < L_i + l_i$, set $\mathcal{F}_i(t-1,j) = \mathcal{F}_i(t,j+1)$ and $\mathcal{P}_i(t-1,j) = j + 1$.

- If $j = L_i + l_i$, then if $\mathcal{F}_i(t,j) < \mathcal{F}_i(t,1) + g_i(\bar{x}_t^i) - \lambda_t\bar{x}_t^i - \mu_t Q_i + h_i(0,1)$, set $\mathcal{F}_i(t-1,j) = \mathcal{F}_i(t,j)$ and $\mathcal{P}_i(t-1,j) = j$; otherwise, $\mathcal{F}_i(t,j) = \mathcal{F}_i(t,1) + g_i(\bar{x}_t^i) - \lambda_t\bar{x}_t^i - \mu_t Q_i + h_i(0,1)$ and $\mathcal{P}_i(t-1,j) = 1$.

Note that $\mathcal{P}_i(t-1,j)$ is the successor index.

2 Set $t \leftarrow t - 1$.
3 If $t > 1$, go to 1.
4 Given the initial condition, j, of the generating unit, the value of $\mathcal{L}_i(\lambda, \mu)$ is $\mathcal{F}_i(1,j)$. Starting from state j at stage 1, trace the successor indices, $\mathcal{P}_i$, to recover an optimal policy.

The previous process can be accelerated by truncating the states at each stage. That is, one can use two states per stage. The first is being on for L_i periods or more and the second is being off for l_i periods or more. We refer the reader to Takriti *et al.* (1996) for more details.

3.2. Initializing and updating the multipliers

The choice of the initial Lagrange multipliers is important since it affects the number of iterations, and hence the execution time. We approximate the production cost – start-up and generation – of a generating unit by a linear function. The slope of this function is computed as $(g_i(Q_i) + h_i(0, 1))/Q_i$. The generating units are ranked in the ascending order of their slopes. For each time period t, we fulfill the demand at a minimal cost. The slope of the last unit to be used is used as an initial value for λ_t and μ_t. This approximation worked quite well with our electric system.

To update the values of the multipliers, we rely on sub-gradient information. Note that

$$\partial \mathcal{L}/\partial \lambda_t = d_t - \sum_{i=1}^{I} x_t^i \text{ and that } \partial \mathcal{L}/\partial \mu_t = d_t + r_t - \sum_{i=1}^{I} Q_i u_t^i.$$

Given a set of multipliers, λ and μ, we evaluate the Lagrangian at three points: (λ, μ), $(\lambda, \mu) + \alpha(\partial \mathcal{L}/\partial \lambda, \partial \mathcal{L}/\partial \mu)$, and $(\lambda, \mu) + \beta(\partial \mathcal{L}/\partial \lambda, \partial \mathcal{L}/\partial \mu)$, where α and β are two distinct positive parameters. We use $\alpha = 1$ and $\beta = 2$ in our computer implementation. Using these three points, one can fit a quadratic function to the Lagrangian and compute the step length for which $\mathcal{L}$ is maximized. This approach outperformed other techniques – sub-gradient and bundle methods – when applied to our test problems. It is important for the reader to realize that updating the multipliers is the most important element in this approach, and that the choice of the updating mechanism should be carefully investigated depending on the electric system under consideration.

3.3. Recovering a feasible solution

The optimization problem of (15.1) is an MIP. As a result, an optimal solution to the dual problem may not possess a corresponding primal feasible solution. To recover feasibility, we adopt the technique of Zhuang and Galiana (1988). Note that a solution is feasible if the spinning-reserve requirements are met. The approach suggested in Zhuang and Galiana (1988) enforces feasibility by increasing the value of the dual multiplier, μ_t, associated with the period with the largest spinning-reserve deficit. The dynamic programming problems are resolved using the new multipliers. Using the new solution, feasibility is checked again and, if infeasible, the multiplier μ_t associated with the period with the largest shortage of reserve is increased. The process is repeated until a feasible solution is achieved. The amount by which a multiplier is increased is a function of the spinning-reserve shortage.

After reaching a feasible solution, the amount of power generated on each unit is recalculated. This is done by fixing the binary variables, u_t^i, and solving the quadratic program

$$\left\{ \min_x \sum_{i=1}^{I} \sum_{t=1}^{T} g_i(x_t^i) : \sum_{i=1}^{I} x_t^i \geq d_t,\ t = 1,\ldots,T, x \in X(u) \right\}$$

The problem of determining an optimal x is called the economic power-dispatch (Frauendorfer *et al.*, 1993; Wood and Wollenberg, 1996). It can be solved easily using a quadratic programming technique.

IV. STOCHASTIC FORMULATION

The formulation of (15.1) assumes that the electric load is known in advance. In reality, the load on the system is a continuous random variable. The value of d_t in (15.1) represents its expected value at t while

$d_t + r_t$ is an estimation of the largest load that may occur. The problem with the formulation of (15.1) rests in the mediation of loads which may result in schedules that are not robust – in other words, schedules that are optimal under d_t but may perform unacceptably poorly under certain load scenarios.

The importance of incorporating load uncertainty into the unit commitment is well recognized (Carpentier *et al.*, 1996; Takriti, *et al.*, 1996) and the subject is quickly becoming a critical area of research. This situation is a result of deregulating the electric power industry: instead of an electric utility controlling the market in a specific region, energy suppliers will be allowed to compete in an open market, and customers will choose their providers. In return, generating companies will have to attract customers by reducing their cost. Therefore, the electric load on the generating system of a utility will suffer from higher demand volatility. We briefly describe how to extend the formulation in (15.1) to include demand uncertainty.

To incorporate the randomness of load into the picture, we use a set of demand vectors or scenarios (Figure 15.2). Each scenario represents a limited amount of information about demand uncertainty. We denote the total number of scenarios in the model by S and the electric load associated with each scenario by d^s, $s = 1, \ldots, S$, where d^s is a vector of size T. Each scenario, s, is assigned a probability, p_s, that reflects the likelihood of its occurrence. That is, we approximate the continuous stochastic load using a number of deterministic realizations. The interested reader may refer to Birge and Louveaux (1997) for more details on stochastic modeling.

Using their load, scenarios can be organized in the form of a tree. Figure 15.2 is an example of a scenario tree with $S = 3$ and $T = 6$. The

Figure 15.2   Example of a scenario tree

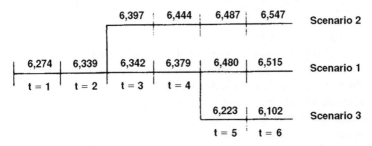

electric load under each of the scenarios is $d^1 = (6274, 6339, 6342, 6379,$
$6480, 6515)$, $d^2 = (6274, 6339, 6397, 6444, 6487, 6547)$, and $d^3 = (6274,$
$6339, 6342, 6379, 6223, 6102)$. To maintain the notations of section III,
we denote the vector of binary decisions for generator i by u^i. It is
understood that u^i consists of a collection of vectors $u^{s,i} \in \{0, 1\}^T$, $s =$
$1, \ldots, S$. Similarly, the amount of power generated, x^i, is the vector of
$x_t^{s,i}$, $t = 1, \ldots, T$, $s = 1, \ldots, S$. Note that if two scenarios, s_1 and s_2,
share the same load prior to t, then the decisions made for them must
be the same. Mathematically, if $d_\tau^{s_1} = d_\tau^{s_2}$, $\tau = 1, \ldots, t$, then

$$u_t^{s_1,i} = u_t^{s_2,i} \text{ and } x_t^{s_1,i} = x_t^{s_2,i}, i = 1, \ldots, I.$$

The previous condition is known as nonanticipativity (Birge and
Louveaux, 1997; Rockafellar and Wets, 1991). It is simply the requirement of a one-to-one correspondence between the scenario tree and
the tree formed by the decisions $(u^{s,i}, x^{s,i})$. We assume that the
nonanticipativity requirement is embedded in the conditions $u^i \in U_i$
and $x^i \in X_i(u^i)$, $i = 1, \ldots, I$.

The objective of the stochastic unit commitment problem is to minimize the expected cost of running the electric system while meeting the
electric load d^s, $s = 1, \ldots, S$. That is

$$\left\{ \min_{u,x} \sum_{i=1}^{I} \sum_{s=1}^{S} p_s \left[\sum_{t=1}^{T} [g_i(x_t^{s,i})u_t^{s,i} + h_i(u_{t-1}^{s,i}, u_t^{s,i})] \right] : \right.$$
$$\left. \sum_{i=1}^{I} x_t^{s,i} \geq d_t^s, \sum_{i=1}^{I} Q_i u_t^{s,i} \geq d_t^s + r_t, u \in U, x \in X(u) \right\}. \quad (15.2)$$

To solve (15.2), we adopt the same approach used in solving (15.1). We
decompose the problem by relaxing the constraints

$$\sum_{i=1}^{I} x_t^{s,i} \geq d_t^s, \sum_{i=1}^{I} Q_i u_t^{s,i} \geq d_t^s + r_t, t = 1, \ldots, T, s = 1, \ldots, S.$$

Note that each scenario has a set of multipliers, λ_t^s and μ_t^s, $t = 1, \ldots, T$.
The resulting single-generator problems are solved recursively using
stochastic dynamic programming. Other aspects of the algorithm described in section III remain unchanged.

The use of dynamic programming in the stochastic case is clarified
through the scenario tree of Figure 15.2. The minimum up- and
minimum down-time are assumed to be 2 and 1, respectively. We solve
one branch at a time – i.e. we start from a certain point in the tree and

Figure 15.3   Dynamic programming applied to the problem of Figure 15.2

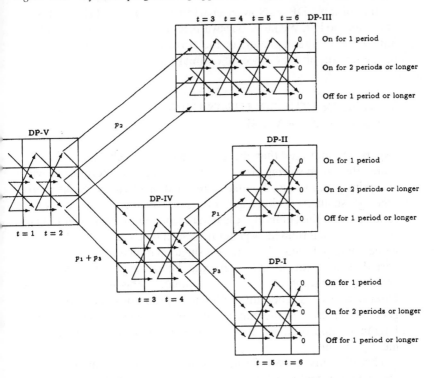

move backward until a branching point is reached. The calculation steps are depicted in Figure 15.3. First, we start by the branch of the last scenario, $s = 3$. That is, we solve the problem defined by Scenario 3 in periods 6 and 5. A sketch of the calculations is demonstrated in the lower-right table labeled DP-I. Then, the dynamic program of $s = 1$ is solved for these two periods – i.e., $t = 6$ and $t = 5$. This is displayed in table DP-II of Figure 15.3. The contents of the first columns in tables DP-I and DP-II are combined by taking the mathematical expectations. The resulting values represent the expected cost-to-go at the end of period 4. The results are copied to the last column of table DP-IV. At this point, one can think of Scenario 1 as having a probability of $p_1 + p_3$. Next, the dynamic program of $s = 2$ is solved. This is displayed in table DP-III in Figure 15.3. After solving for Scenario 2, we move to Scenario 1 and solve its dynamic program. Note that we need to solve only for periods $t = 4$ and $t = 3$. The calculations are performed in table DP-IV. By taking the mathematical expectation over the first column of table

DP-III and DP-IV, we obtain the values of the entries in the last column of table DP-V. Now, Scenario 1 has a probability 1 and is the only scenario left. We move backward, as demonstrated by table DP-V, until we reach the beginning of the horizon. Given the initial state of the generator, one can trace an optimal solution.

V. REFINING THE SOLUTION USING INTEGER PROGRAMMING

Experience indicates that the Lagrangian approach of section III is quite effective in solving (15.1). The solution process terminates when an optimal solution to the dual is reached or when the duality gap is relatively small. At period t, generators for which $u_t^i = 1$ and $x_t^i < Q_i$ are called the marginal units at t. Clearly, the marginal cost of operating the system, λ_t, is determined by generators such that $q_i \leq x_t^i < Q_i$, hence the name 'marginal units'. If there is a single marginal unit for each time period, then the duality gap is zero and the Lagrangian approach yields an optimal solution to (15.1). Unfortunately, this is rarely the case. Often, a utility has generators with similar properties. As the number of generators increases, the number of marginal units also grows and the Lagrangian approach becomes less effective in solving the unit commitment problem.

This difficulty is best clarified using an example. Suppose that a generating system consists of two identical units. Let us also assume that our planning horizon is one period, $T = 1$, and that the electric load, d_1, is equal to the minimum operating capacity, q. Since both generators have the same cost function, the Lagrangian solution method would result in $u_1^1 = u_1^2$ and $x_1^1 = x_1^2$. That is, both units will be switched on and off simultaneously. Furthermore, when both are on, the amount of power generated is at least $2q$. An optimal solution in this case would be to switch one unit on and to turn the other off, a solution which cannot be found by Lagrangian relaxation.

The problem of having similar generating units is becoming increasingly important as a result of deregulation. Only competitive and effective generators will survive in the new market. Utilities – generating companies – are investing in smaller, more economic, and less polluting gas turbines. Although manufacturers of such generators may differ, most of these units have comparable cost functions. Another complicating factor in a deregulated environment is the notion of electricity contracts. A contract that offers electric power can be modeled by the buyer as a generating unit. Given that a contract must be competitive if compared to others available in the market, there is no doubt that using the dual approach may result in decisions that are far from being optimal.

To handle the difficulty of having a number of generators – contracts – on margin, we make use of the schedules, u^i, generated during the different iterations of the approach of Section III. Numerical experience indicates that after a few iterations, the resulting schedules, u^i, for each generating unit tend to repeat. In other words, in the first few iterations, a set of feasible schedules for each generator is produced. Later on, the new values of λ and μ do not produce new schedules. Instead, they try to find an optimal combination of the existing feasible schedules. Even though new schedules may be obtained in later iterations, their contribution to the quality of the solution is minimal. This is the main idea behind our approach. Rather than aiming for an optimal solution for the dual by performing a large number of iterations, one can use the schedules produced after the first few iterations of the algorithm to construct a solution for the unit commitment problem.

The previous idea is motivated by the proof of section VI in Bertsekas *et al.* (1983). Although the result of Bertsekas *et al.* (1983) is established for the case in which g_i is piecewise-linear, we extend this result to the case in which g_i is quadratic. The goal behind the following presentation is to familiarize the reader with the idea of Bertsekas *et al.* (1983) and to prepare the ground for a smooth transition into our refinement approach.

Theorem 1 Assume that $I > 2T$. Then, the difference between the optimal value of (15.1) and the optimal value of $\mathcal{L}(\lambda, \mu)$ is bounded above by $2T(h^* + Tg^*)$, where $h^* = \max_i h_i(0,1)$ and $g^* = \max_i g_i(q_i)$.

Proof For each unit i, select m distinct points within the range $[q_i, Q_i]$. Let us enumerate all feasible schedules – binary sequences $u^i \in U_i$ – for the generating units, $i = 1, \ldots, I$. For each schedule u^i, we create a set of sequences x^i. This is done by setting x^i_t to q_i, Q_i, and each of the m selected points if $u^i_t = 1$. As a result, each generator i has K_i generating schedules which we denote by $x^i(k)$, $k = 1, \ldots, K_i$. Note that the cost of a generating schedule, $x^i(k)$, is $S_i(k) + \Sigma_{t=1}^T g_i(x^i_t(k))u^i_t(k)$, where $S_i(k)$ is the total start-up cost associated with this schedule. That is, $S_i(k) = \Sigma_{t=1}^T h_i(u^i_{t-1}(k), u^i_t(k))$.

Consider the linear program

$$\min_v \sum_{i=1}^{I} \sum_{k=1}^{K_i} \left[S_i(k) + \sum_{t=1}^{T} g_i(x_t^i(k)) u_t^i(k) \right] v_i(k)$$

$$\text{s.t.} \quad \sum_{i=1}^{I} \sum_{k=1}^{K_i} x_t^i(k) v_i(k) \geq d_t, \, t = 1, \ldots, T,$$

$$\sum_{i=1}^{I} \sum_{k=1}^{K_i} Q_i u_t^i(k) v_i(k) \geq d_t + r_t, t = 1, \ldots, T,$$

$$\sum_{k=1}^{K_i} v_i(k) = 1, i = 1 \ldots, I, \quad (15.3)$$

where $v_i(k) \geq 0$. In a basic solution to (15.3), there are at most $2T + I$ positive variables. The constraints $\Sigma_{k=1}^{k} v_i(k) = 1$, $i = 1, \ldots, I$, require that at least I variables be positive. As a result, the remaining $2T$ variables are associated with at most $2T$ generators. Hence, $I - 2T$ generators are associated with integer decisions $v_i(k)$. To construct an integer solution to (15.3), each of the $2T$ generators which are associated with noninteger $v_i(k)$ is switched on throughout the planning horizon. These units are operated at their minimum capacities, q_i, until needed. Then, the cost of constructing an integer solution for (15.3) is at most $2T(h^* + Tg^*)$, where $h^* = \max_i h_i(0, 1)$ and $g^* = \max_i g_i(q_i)$. As m increases, the optimal value of (15.3) provides a lower bound on $\mathcal{L}$ and the integer solution of (15.3) provides an upper bound on (15.1). Hence, the duality gap of the solution approach of section III is bounded above by $2T(h^* + Tg^*)$. ∎

The previous proof enumerates all possible generating schedules and uses them to construct an upper bound on the duality gap. We suggest using the binary schedules generated during the Lagrangian iterations to improve the quality of the solution.

We denote the number of dual iterations by K. At each iteration, $k = 1, \ldots, K$, we use the Lagrange multipliers, $\lambda(k)$ and $\mu(k)$, to obtain a solution, $u(k)$ and $x(k)$, that indicates the status of a unit and the load on it. As we perform more iterations, we hope that the values of $u(k)$ and $x(k)$ are getting closer to an optimal solution. Our approach uses the vectors $u(k) \in U$ to construct a feasible solution for (15.1). To do so, we associate a binary variable, $v_i(k)$, with each vector $u^i(k)$. Our goal is to select for each generator, i, a schedule, $u^i(k)$, so that all unit commitment constraints are met. Note that any $u^i(k)$ satisfies the minimum up- and down-time requirements for i. Therefore, the only constraints we need to worry about are satisfying the electric load, spinning reserves, and the operating capacity of a unit.

Mathematically, the suggested model finds an optimal unit commitment by solving the integer program

$$\min_{x,v} \sum_{i=1}^{I} \sum_{k=1}^{K} \left[S_i(k) + \sum_{t=1}^{T} g_i(x_t^i(k)) u_t^i(k) \right] v_i(k)$$

$$\text{s.t.} \quad \sum_{i=1}^{I} \sum_{k=1}^{K} x_t^i(k) v_i(k) \geq d_t + r_t, \ t = 1, \ldots, T,$$

$$\sum_{i=1}^{I} \sum_{k=1}^{K} Q_i u_t^i(k) v_i(k) \geq d_t + \quad t = 1, \ldots, T,$$

$$\sum_{k=1}^{K} v_i(k) = 1, \ i = 1 \ldots, I,$$

$$q_i u_t^i(k) v_i(k) \leq x_t^i(k) \leq Q_i u_t^i(k) v_i(k), \ k = 1, \ldots, K,$$

$$t = 1, \ldots, T, \ i = 1, \ldots, I. \tag{15.4}$$

Here, $v_i(k)$ is a binary variable. The constraint $\Sigma_{k=1}^{k} v_i(k) = 1$ indicates that one and only one schedule needs to be chosen for a generator, i. Note that the minimization in (15.4) is over x and v in contrast to (15.3) where the minimization is over v only. To use branch-and-bound on (15.4) effectively, its continuous relaxation must be easy to solve. This is not the case in our model: there are elements in the objective function in which decision variables, $x_t^i(k)$ and $v_i(k)$, are multiplied. As a result, the relaxed model is nonconvex which makes an optimal solution difficult to find. To avoid this problem, we reformulate (15.4) as

$$\min_{x,y,v,z} \sum_{i=1}^{I} z_i$$

$$\text{s.t.} \quad \sum_{i=1}^{I} \sum_{k=1}^{K} x_t^i(k) \geq d_t, \ t = 1, \ldots, T,$$

$$\sum_{i=1}^{I} \sum_{k=1}^{K} Q_i u_t^i(k) v_i(k) \geq d_t + r_t, t = 1, \ldots, T,$$

$$\sum_{k=1}^{K} v_i(k) = 1, \ i = 1 \ldots, I,$$

$$q_i u_t^i(k) v_i(k) \leq x_t^i(k) \leq Q_i u_t^i(k) v_i(k), \ k = 1, \ldots, K,$$

$$t = 1, \ldots, T, \ i = 1, \ldots, I$$

$$\sum_{t=1}^{T} g_i(x_t^i(k)) u_t^i(k) \leq z_i + y_i(k) - S_i(k), \ k = 1, \ldots, K,$$

$$i = 1, \ldots, I$$

$$0 \leq y_i(k) \leq M(1 - v_i(k)), \ k = 1, \ldots, K, \ i = 1, \ldots, I. \tag{15.5}$$

where M is a sufficiently large number. Note that we added two variables: z and y. The value of z_i, $i = 1, \ldots, I$, represents the cost of running plant i under the selected schedule. If $v_i(k) = 0$, then $y_i(k)$ can take positive values which guarantees the feasibility of $\sum_{t=1}^{T} g_i(x_t^i(k))u_t^i(k) \leq z_i + y_i(k) - S_i(k)$. When $v_i(k) = 1$, then $y_i(k) = 0$ and z_i is set to $S_i(k) + \sum_{t=1}^{T} g_i(x_t^i(k))u_t^i(k)$.

It is clear that if $v_i(k) = 0$, then the corresponding x variables do not play a role in the optimization as $y_i(k)$ offsets their role. To reduce the number of variables in the model, we reformulate (15.5) as

$$\min_{x,y,v,z} \quad \sum_{i=1}^{I} z_i$$

$$\text{s.t.} \quad \sum_{i=1}^{I} x_t^i \geq d_t, \ t = 1, \ldots, T,$$

$$\sum_{i=1}^{I} \sum_{k=1}^{K} Q_i u_t^i(k) v_i(k) \geq d_t + r_t, t = 1, \ldots, T,$$

$$\sum_{k=1}^{K} v_i(k) = 1, \ i = 1 \ldots, I,$$

$$q_i \sum_{k=1}^{K} u_t^i(k) v_i(k) \leq x_t^i \leq Q_i \sum_{k=1}^{K} u_t^i(k) v_i(k), \ t = 1, \ldots, T, \ i = 1, \ldots, I,$$

$$\sum_{t=1}^{T} g_i(x_t^i) u_t^i(k) \leq z_i + y_i(k) - S_i(k), \ k = 1, \ldots, K, \ i = 1, \ldots, I,$$

$$0 \leq y_i(k) \leq M(1 - v_i(k)), \ k = 1, \ldots, K, \ i = 1, \ldots, I.$$
$$(15.6)$$

The program in (15.6) is an MIP of a smaller size than (15.1). It can be solved using branch-and-bound. If we relax the integer requirements on v, the resulting problem is a convex mathematical program. Furthermore, for a given v, program (15.6) is an economic power-dispatch. If feasible, the optimal value of (15.6) is an upper bound on the optimal value of the unit commitment problem. Note that the solution of (15.1) obtained from solving the dual corresponds to the solution of (15.6) when $v_i(K)$ is set to 1 for all $i = 1, \ldots, I$. As a result, the optimal value of (15.6) is guaranteed to be no worse than the best solution obtained using the dual approach of section III.

VI. NUMERICAL RESULTS

We applied the solution approach of section III and section V to a generating system based on the system of an electric utility in the Midwestern United States. The duration of the study horizon was 1 week which was split into $T = 168$ hours. The number of generating units was changed to study its correlation with the improvement that resulted from using the refinement technique of section V. The

Lagrangian-relaxation approach was implemented using the C language with considerable effort put into optimizing the code and tuning its performance. The MIP of (15.6) was solved using branch-and-bound. Table 15.1 presents the results of our tests. The first column provides the number of generators n. The generating units were split into sets of similar generators. For example, $n = 2 \times 3$ indicates that there are 2 sets of generators each of which contains 3 identical generating units. Owing to the use of different number of generators, the electric load, d, was changed to correspond to the total generating capacity of the system. The column labeled 'K' contains the number of Lagrangian iterations performed. To study the effect of K, we stopped the Lagrangian iteration process after $K = 30$ and $K = 100$ iterations. For the Lagrangian-relaxation approach, the best value of the Lagrangian is provided in the column entitled "Dual", the primal solu-

Table 15.1   Lagrangian-relaxation of section III vs. refinement approach of section V

n	Lagrangian-relaxation					Refinement		
	K	Primal	Dual	Gap %	CPU	Primal	Gap %	CPU
1×2	30	619543	525619	17.87	0.2	593421	12.90	4.3
	100	619543	556446	11.34	0.6	593421	6.64	9.0
1×3	30	945797	837204	12.97	1.0	889470	6.24	52.4
	100	945797	843954	12.07	1.4	889470	5.39	75.8
1×4	30	1265867	1155635	9.54	1.2	1184765	2.52	148.2
	100	1265867	1158330	9.28	1.8	1184765	2.28	245.1
1×5	30	1588537	1463758	8.52	1.6	1483221	1.33	483.2
	100	1588537	1466200	8.34	2.3	1480956	1.01	766.6
2×2	30	1357791	1239506	9.54	1.3	1271270	2.56	12.5
	100	1357791	1243822	9.16	1.3	1271076	2.19	20.5
2×3	30	2050403	1892517	8.34	1.99	1906501	0.74	27.8
	100	2050403	1894769	8.21	2.76	1906115	0.60	84.9
2×4	30	2742597	2532558	8.29	2.37	2542539	0.39	66.8
	100	2742597	2534191	8.22	3.59	2542153	0.31	91.7
2×5	30	3433783	3171674	8.26	2.98	3177771	0.19	84.5
	100	3433783	3173569	8.20	4.39	3177191	0.11	117.4
3×2	30	1996820	1883180	6.03	0.5	1893508	0.55	73.2
	100	1996820	1884839	5.94	1.1	1893508	0.46	113.4
3×3	30	3007650	2838669	5.95	0.7	2840261	0.06	167.1
	100	3007650	2839550	5.92	2.1	2840186	0.02	291.6
3×4	30	4019094	3786903	6.13	1.1	3787506	0.02	353.8
	100	4019094	3787500	6.11	2.7	3787506	0.00	510.6
3×5	30	5027725	4732773	6.23	1.3	4734159	0.03	463.8
	100	5027725	4732837	6.23	3.4	4733834	0.02	526.2

tion corresponding to it in the column "Primal", and the duality gap measured as a percentage of the dual solution. For example, for the case in which $n = 3 \times 2$ and $K = 30$, the duality gap is computed as $(1\,996\,820 - 1\,883\,180)/1\,883\,180 = 6.03\%$. The 'CPU' column provides the execution time measured on an IBM RS/6000 running AIX 4.1 with a 120 MHz processor and 250 MBytes of memory. The column labeled 'Primal' under 'Refinement' provides the best solution found using the refinement approach of section V. The duality gap for the refined solution is computed against the dual provided by the Lagrangian.

As the number of iterations K increased from 30 to 100, the value of the primal solutions obtained using the Lagrangian-relaxation approach remained the same. However, as K increased, the value of the dual improved. That is, the algorithm reached a good generating schedule at an early stage but needed a large number of iterations to establish a good lower bound. The primal solution obtained from the refinement approach is significantly better than the primal solution of the 'Lagrangian-relaxation.' Furthermore, the quality of the refined solution improves as the number of marginal generators increases. For example, when $k = 30$, if the system has 3×2 units, the error of the refined solution is 0.55% compared to 6.03% for the Lagrangian approach. When the system has 3×3 generators, the error in the refined solution becomes 0.06% compared to 5.95% for the Lagrangian approach.

It is worthwhile mentioning that the primal solution of the refinement approach was almost the same whether K is set to 30 or 100. This is a result of the generating schedules, $u(k)$, being duplicated between different iterations. To achieve good performance, one may consider passing the first few schedules to the refinement approach and iterate using the Lagrangian method to obtain a good lower bound.

VII. CONCLUSION

We presented a solution method for the unit commitment problem for both the deterministic and stochastic cases. The method is based on Lagrangian-relaxation and dynamic programming which are easy to implement. Given that the Lagrangian-relaxation approach may run into difficulties if some of the generating units share the same characteristics, we developed a refinement approach that uses the schedules obtained during the Lagrangian iterations to provide a better primal solution. Numerical experience indicates that the refined solution provides significant improvements in the quality of the solution.

References

Bertsekas, D.P., G.S. Lauer, N.R. Sandell and T.A. Posbergh (1983) 'Optimal Short-term Scheduling of Large-scale Power Systems', *IEEE Transactions on Automatic Control*, 28, 1–11.

Birge J.R. and F. Louveaux (1997) *Introduction to Stochastic Programming* (Berlin: Springer-Verlag).

Carpentier, A., G. Cohen, J.-C. Culioli and A. Renaud (1996) 'Stochastic Optimization of Unit Commitment: A New Decomposition Framework', *IEEE Transactions on Power Systems*, 11, 1067–73.

Cohen, A.I., and M. Yoshimura (1983) 'A Branch and Bound Algorithm for Unit Commitment', *IEEE Transactions on Power Apparatus and Systems*, 102, 444–51.

Frauendorfer, K., H. Glavitsch and R. Bacher (eds) (1993) *Optimization in Planning and Operation of Electric Power Systems* (Berlin: Physica-Verlag).

Hobbs, W.J., G. Hermon, S. Warner and G.B. Sheblé (1987) 'Dynamic Programming Approach to Unit Commitment', *IEEE Transactions on Power Systems*, 2, 339–50.

Kazarlis, S.A., A.G. Bakirtzis and V. Petridis (1996) 'A Genetic Algorithm Solution to the Unit Commitment Problem', *IEEE Transactions on Power Systems*, 11, 83–92.

Lee, F.N. (1988) 'Short-term Thermal Unit Commitment – A New Method', *IEEE Transactions on Power Systems*, 3, 421–8.

Möller, A. and W. Römisch (1995) 'A Dual Method for the Unit Commitment Problem', *Technical Report 95–1* (Berlin: Institut für Mathematik der Humboldt-Universität zu Berlin).

Muckstadt, J.A. and S.A. Koenig (1997) 'An Application of Lagrangian Relaxation to Scheduling in Power-generation Systems', *Operations Research*, 25, 387–403.

Rockafellar, R.T. and R. J.-B. Wets (1991) 'Scenarios and Policy Aggregation in Optimization under Uncertainty', *Mathematics of Operations Research*, 16, 119–47.

Schramm, H. and J. Zowe (1992) 'A Version of the Bundle Idea for Minimizing a Nonsmooth Function: Conceptual Idea, Convergence Analysis, Numerical Results', *SIAM Journal on Optimization*, 2, 121–52.

Takriti, S., J.R. Birge and E. Long (1996) 'A Stochastic Model for the Unit Commitment Problem', *IEEE Transactions on Power Systems* 11, 1497–1508.

Wood, A.J. and B.F. Wollenberg (1996) *Power Generation, Operation, and Control* New York: John Wiley.

Zhuang, F. and F.D. Galiana (1988) 'Towards a More Rigorous and Practical Unit Commitment by Lagrangian Relaxation', *IEEE Transactions on Power Systems*, 3, 763–73.

Zhuang, F. and F.D. Galiana (1990) 'Unit Commitment by Simulated Annealing', *IEEE Transactions on Power Systems*, 5, 311–18.

Abbreviations

AMPL	A Mathematical Programming Language, ILOG
ATV	automated transfer vehicle
BOM	bill of material
BTU	British thermal unit, fuel energy
CIP	corporate inventory point
DAS	demand adaptive system
DT	decision tree
EM	IBM EasyModeler
EMOSL	Entity Modelling and Optimization Subroutine Library, DASH
ESA	European Space Agency
GPI	generalized pairwise interchange
HPPVM	high-performance parallel virtual machine
MEMIPS	Model Enhanced Solution Methods for Integer Programming Software
MCG	multiple classifier generation
MILP	mixed integer linear programming
MINLP	mixed integer nonlinear programming
MOPS	Mathematical Optimization System, Prof. U. Suhl
MPSX	IBM Mathematical Programming System Extended
MRP	material requirement planning
MSG	mail stream generation
MSO	mail stream optimization
MSS	mail stream selection
MSU	mail stream update
MWH	mega Watt hours, generated energy
NN	neural network
NLP	nonlinear programming
OA	outer approximation
OPN	ordering part number
OSL	IBM Optimization Subroutine Library
PVM	parallel virtual machine
RFM	recency, frequency, and monetary value
RM	row material
ROI	return of investment
RPU	reprocessing unit
SKU	stock holding units
SOS	special ordered set
TS	tabu search
TSP	traveling salesman problem
UC	unit commitment
WIP	work in progress

Index

323